MW00827230

Stop
Overspending

Stop Overspending

WHY MOST PEOPLE CAN'T SAVE MONEY

A Closer Look at Financially Destructive Behaviors
and How to Overcome Them

Integrated Wealth Theory™ by:

Buğra Bakan, CFP®

Published by Buğra Bakan.

Library of Congress Cataloging-in-Publication Data
Names: Bakan, Buğra, author.
Title: Stop Overspending / Buğra Bakan.
Cover design by Joan Carlson
Portrait Photography: Mary Small Photography
Manufactured in the United States of America
First Edition

Dedicated to those;

Who have encouraged me to write this book.
I hope it is worthy of your praise.

Preface

If it feels as if you've got a secret hole in your pocket through which your hard-earned dollars slip away, as if no solution you've looked into has been able to stop the bleeding, and as if you're ready to expand your horizon and embrace a wider set of ideas to address this problem — then this is the right book for you. Money is a big idea, so how to make the most of it, must also be a big idea. Just saying that you need to budget, doesn't cut it, especially in the digital world we now live in.

It's as if one day we woke up to a post-modern, globalized, neo-liberal, digitalized, automated, and integrated New World. This may sound like a cliché unless we define what is meant by "new." In hindsight it's easier to identify the old paradigm, because we could be in a transition period, or going through a trial-and-error phase. But even still, we may know enough to say that the old way of doing things, and the institutions that supported it, falls short of answering today's needs. The 18th c. Industrial Revolution-based, socio-cultural, political, and economic institutions, along with their analog operating manuals, are predictably outdated. Digital Revolution-based production, consumption and communication processes, continue to be perpetually new. So there's a disconnect between the old institutions, and the needs of this new age.

The speed of change in the last two to three decades has been a challenge to adapt to, even for Silicon Valley engineers. My generation, those of us in our 40s and older, remember black and white TVs vividly. I remember the day my family got our first color TV. When I came home from school, I immediately knew something important was happening in our house. Today, I carry around a much more advanced TV in my pocket, and it's also a radio, a phone, a library, a GPS, a health monitor, and so on. When I began writing these pages back in 2018, the idea that we're living in a New World had already become very obvious to me.

Now, in a lock-down resulting from the spread of Covid-19, as I put the finishing touches on this book, I watch the developments around the globe and my conviction has only strengthened. The way business is done, the rules of governance, and social contracts everywhere in the world, are going to be torn apart and re-written in the next few decades. We've moved on to a New World, but the institutions, socio-economic and socio-cultural systems which we rely on to organize our lives, have fallen behind. This friction is manifesting itself in popular discontent, and as a result, demagogue politicians are finding the perfect habitat to bloom into power.

In this New World, money rules have changed in tandem with the emergence of the 1980's neo-liberal policies, and it seems that these changes will continue to adapt. This shift was most dramatically solidified during the 2008 Global Financial Crisis. Those who understand these permanent modifications well will adapt, and those who fail to do so, will unfortunately be left behind. With this book, I hope to start a conversation around the best strategies for facing these challenges, especially as they relate to personal finance.

One major shift affecting every modern individual's financial well being, is the significant drop in the safety net provided by private and public-sponsored pension funds. In the "old days," all you had to do was find a good job at a large, reputable company, work there for 30 years, and collect a pension for the rest of your life. That, along with your Social Security retirement benefits, would be enough to sustain a quality of life long past your working years. This was true for both blue and white-collar workers. You could raise a family, your kids would attend public schools, life wasn't too expensive to manage, and your legacy would be passed on. Lastly, your health insurance was included in your total compensation, with only a relatively small amount having to be paid by you, the employee. Today, the responsibility of preparing for retirement falls solely on each person, as pensions are a thing of the past. To a large extent, the same can be said about the availability of free public education and health care insurance.

At the same time, materialistic consumerism has turned into the World's most popular religion. Central banks opened the flood gates for easy access to money via credit, facilitating an ability to purchase anything, anytime. And e-commerce made it easier than ever to shop in the comfort of our own homes.

This is the New World I refer to, one that provides everything a person needs to help them consume incessantly. It's also one that offers no help, direction, or incentive to save — and the dire consequences of this are treated like a hot potato. The rug is eventually pulled out from under the individual, and for the most part, they're unaware of what's about to hit them. We'll see vast numbers of people, victims of high levels of debt falling on their faces in the next few decades, and looming over it all, the environmental consequences of this consumption frenzy is casting its dark shadow. If we can't stop it, we should at least be prepared for this crash. We see that our policy agenda is instead busy chasing populist demands.

Over the course of 20 plus years in my career in financial services, I've heard the following phrases hundreds, if not thousands of times:

> "I work hard, do all that I can, and in the end, I have nothing to show for it."

> "We make a decent living, but have no money left in the bank — where does the money go?"

Earlier in my career, my answer to these complaints involved providing budgeting sheets and sending clients articles on how to be frugal, only to see no progress. I started to recognize a familiar and troubling pattern everywhere I looked. Honest, hardworking people, while doing their part, were still having a hard time building a nest egg. My budgeting proposals, in some ways, made things worse for them, as many simply got the idea that being irresponsible was the sole reason for their financial demise. I couldn't help but wonder, is there more to this problem than meets the eye?

Once you hang a question on your subconscious thought hanger, in the back of your mind even when you're not aware of it, you're working on an answer and continue to search for what makes the most sense. I researched and collected every bit of information I could get my hands on to answer the question: how can I do a better job of helping these folks who are struggling to keep their finances in order? As my usual solution wasn't working for them, and I wasn't able to find anything else more effective, I realized I had to create a path of my own. Stop Overspending and the Integrated Wealth Theory™ as put forward in these pages, is the result.

I concluded that telling a person to be frugal addresses only a very small part of the problem, and that's why it failed so often. A more comprehensive remedy was needed because we are, for the most part, a product of our environment. Without an understanding of that environment and how it affects our behavior, my proposals would be useless.

This then, is what I am aiming to do with this book. My goal is to convey the big picture, and how people can make the right changes to fit into it. Arming them with this new information, I intend that my readers can gain a new perspective, that will help them organize and finance their life goals more effectively.

I employ a variety of disciplines to address these complex issues, and as you read along, you'll see that seemingly unrelated concepts will begin to come together like pieces of a jigsaw puzzle. I ask you the reader to be patient, for the unified solution will reveal itself. The progression of my logic is:

1) Understand the external environment we live in and its role in our consumption decisions.

2) Understand our biological structure and its impact on our thoughts, feelings, and emotions, especially as related to our consumption decisions.

3) Look for ways to use this information to improve decision-making.

4) Present the techniques available for this purpose.

5) Introduce practical information to better organize finances and investments.

6) Examine influential thought leaders' theories and practices on finding happiness and living a good life.

7) Connect the dots and absorb the concepts of the complete picture with Integrated Wealth Theory.™

Over the past few years, to help my clients and friends save and invest, and to help them manage their life goals more effectively, I've been sharing bits of this information with them. The response is usually: "You should put all this in a book, people should know about it." With that encouragement, I decided to compile and present it in this more formal and better-organized form. My hope is that after reading this book, the reader will not only have a better understanding of the internal and external factors which lie at the root of their difficulty with managing personal finances, but also I intend for the solutions presented to serve as a pragmatic guide to taking prescribed steps toward their ultimate progress and success.

Pulling from my decades of observations in the field, I've identified two new mental biases which I submit as additions to the current literature on hand in behavioral finance: the *Lazy Option Bias*, and the *Opportunity Cost Blindness*. As you read about them throughout this book, I hope you'll find these to be valuable contributions.

With the right mindset leading to right actions, change can be instant and profound. If I can empower the reader even slightly to be better at organizing their finances, finding happiness, and living a good life by reaching their life goals, my time has been well spent, and this book has done its job. It's my sincere wish that you'll enjoy reading it, and that you implement the information presented here to better your life. Before we start, let's set the tone and get in the right mindset with one of my favorite quotes from Goethe:

**I have come to the frightening conclusion that
I am the decisive element.**

It is my personal approach that creates the climate.

It is my daily mood that makes the weather.

Contents

STOP OVERSPENDING

1
Introduction

According to the United States Census Bureau,[1] only a third of Americans are contributing to a workplace retirement account. In a survey conducted by GOBankingRates,[2] 42% have less than $10,000 in a savings account, 64% will retire broke, and to add insult to injury, the NY Federal Reserve Bank (FED), reported that household debt reached $14.3 trillion in the first quarter of 2020.[3] According to the FED, the average household is carrying $16,000 in credit card debt.

The words above were written prior to the Coronavirus pandemic upon us currently. After being hit with Covid-19, we now know that most Americans are living paycheck to paycheck and don't have any money saved for emergencies. The financial status of the citizens of the strongest country in the world is in dire shape.

How did we end up here? Do we have extremely low income per capita in the United States of America, or a terrible education system? Does our economy produce low value-added goods and services? Are we lacking competitiveness in the international markets?

On the contrary, US Gross Domestic Product (GDP) tops over $20 trillion[4] and income per capita is around $55,000. The US is a world leader in high-value-added technological innovation, our goods and services are competitive, and the working-age population rate is uniquely high compared to other countries in the developed world. We have the unmatched advantage of having deep and efficient financial markets to support the functioning of our economic system.

Then, is it because most people are unaware that to reach their life goals they need to live below their means, proactively saving and investing along the way toward those goals? I seriously doubt that. It's fair to assume that most individuals in the US who've engaged in various economic transactions of one kind or another, understand this very basic fact of modern life. But then, why do they fail in execution? What is the root of this disconnect?

In this book, we'll explore many similar questions, studying both background information and likely answers. If you feel like you don't have full control of your finances; if you know you should be saving and investing more; if you find yourself always paddling upstream and you're exhausted by the rat race — if you're looking for a fresh approach and a discerning voice to guide you to better solutions — this book is for you. My purpose is to address these concerns, challenge this complex problem, and answer the following questions:

- What aspects of our social, economic and political structures are creating roadblocks to an individual's path to financial freedom?

- How does our brain biology and biochemistry contribute to our financial demise?

- Can a person work toward inner peace and financial stability at the same time? Are these two goals mutually exclusive?

- What can be done to improve the financial balance sheets of American citizens?

In the chapters ahead, I'll share my take on why our society fails to deliver for Main Street folks, while helping Wall Street investors get ever-wealthier. I'll present some provocative ideas about our current and popular political views, socio-economic and governmental institutions, including our educational system, and will also share information on brain biology, which has some very interesting insights to impart.

Ultimately, I suggest that the dire situation at hand can only be remedied by a concerted effort by all parties, regardless of how unrelated they may seem — hence the name for my thought and practice technology: Integrated Wealth Theory.™

We will simply not be able to solve this problem effectively unless we get a better understanding of:

- Human brain chemistry

- The affect that hormones have on our feelings, emotions, and actions

- Our psychological responses to pain, pleasure, and rewards

- Socio-political and economic environments we live in

- The rules of personal finance.

This by no means absolves the individual from personal responsibility, it just advises the state, corporations, non-profits, and our education system to provide people with the necessary tools to create lives that are financially stable and meaningful, while looking for ways to empower each person to rise to the task as a matter of American culture. Later, we'll see that according to Aristotle, this cultural education should fall on the shoulders of the State, so I'm not in fact proposing that we follow a brand new idea.

For this book to add value and capture your imagination, I invite you, the reader, to have an open mind to alternative views, to challenges of the status quo, and to preconceived notions. It's a well-known analogy that aptly states: the mind is just like a parachute, it only functions when it's open.

Many people believe that the modern era's most common problem is depression. I suggest that instead it may be an obsession with a loosely-defined concept called "happiness." As a result, many of us spend our lives chasing this imaginary rabbit, which results in nothing but dissatisfaction. This never-ending chase has been arguably instrumental in costing us our retirement funds, our kids' college savings, and myriad other financial aspirations. Very few people can define what happiness truly means for them, and yet, most would describe it as a primary goal in life. Toward the end of this book, I invite the insights of world-renowned philosophers to this discussion, so we can take a critical look at what happiness actually is, and how to build a good life. We'll see that no sage has ever advised finding happiness in overconsumption — though so many head in the opposite direction on this point.

We're constantly bombarded with marketing messages that masterfully convince us of all the things we need to buy, with money we don't have, to impress people we don't even know. Comedian George Carlin nailed it in his skit titled: "My Stuff." We've built a society of consumers that marketing firms, and the corporations that hire them, know exactly how to manipulate. These corporations have mounds of detailed information about us, which they then use to corral us into ways of thinking and precise behaviors, which suit *their* needs and boost *their* profits. With technology's help, they get better and better at this, by using the data we provide to them through our web purchases and social media presence.

Could it be true that the socio-economic root cause of this predicament, (that is, the success of the individualistic capitalism model — which helped the US become the mighty power it is in the first place), is now paradoxically bringing about the financial and spiritual demise of the majority of its citizenry?

Almost all our political and economic systems and institutions are the fruits of the 18th c. Industrial Revolution. Capitalism, communism, liberalism, individualism, consumerism, and nationalism were all shaped primarily by that era's thinkers and events. The problem is, we live in a vastly different world today, and there's a disconnect between everyday conditions and the institutions that are designed to organize the society around those conditions. While the rich get (significantly) richer, more and more people are being (or feeling) left behind. This massive inequality creates a disgruntled sentiment, which is threatening the stability of the systems we rely on to function as a stable, global society. As a result, topping the list of potential threats, is the rise of populist autocratic governments that we find spreading around the globe.

When did it become acceptable for the richest one percent to own more wealth than the bottom 90 percent? How is this supposed to be OK? In an interview on CNBC, Ray Dalio, the manager of the world's largest hedge fund, Blackwater Associates, called out that the low-interest rates and quantitative easing policies rolled out by central banks following the 2008 Global Financial Crisis, were the root cause of this monumental wealth gap and the global populist waves which have grown from it.

Populism empowers irrational, shortsighted public policies and erodes the very institutions that are designed to benefit the people, even those now demanding extreme measures from today's politicians. This trend paves the way to polarization, hurts our economy, our democracy, and slows down the natural progress of moving toward a more stable society. When the macro environment becomes less stable, your own (financial) stability becomes even more important.

It's almost impossible to isolate one's self from these macro trends. According to Harry Stack Sullivan[5] (1892-1949), a Neo-Freudian psychiatrist and psychoanalyst: an individual's life is an *expression of complex social networks and interpersonal relationships.* George Herbert Mead [6] (1863-1931), a philosopher, sociologist, and psychologist who helped form the concept of "symbolic interactionism," described the individual as "the interiorized other," meaning *the aggregated sum of others' thoughts and expressions of ourselves.* A Sufi tradition teaches us: *"We hold a mirror to one another."* In this idea, we learn that we are not only reactionary beings, but as social animals as well, that we exist as integral and interactive parts of our society.

We're a product of our environment, true, but the story doesn't end there. We also have to consider the role played by our nature, brain structure, psychological state, and how each of these affect and contribute to our challenges in life.

In Western countries, the usual modus operandi or status quo is to be individualistic and competitive. We are conditioned to judge our success in comparison to others, and most often success is measured by financial net worth.

There is now a great amount of literature in the relatively new field of behavioral finance, showing how unhelpful or flat out self-destructive our thought processes and choices can be when it comes to our decisions around money and personal finance. If our goal is to produce sustainable and lasting solutions, we need to understand the roots of these counter-productive behaviors and be able to offer alternatives.

There's also been a large body of wisdom passed on to us across generations, going back thousands of years now, on how to improve our mental state and overall life satisfaction through mindful awareness and meditation.

And finally, many great minds have contemplated the topic of what makes the best form of governance. Whether it's better to employ individual-worshiping capitalism or state-worshiping communism, monarchy versus democracy, or countless other possibilities. I researched the stances taken by these thinkers to help us find solutions from a macro and historical perspective. As I've studied politics, history, and philosophy, I've come to realize that people actually change very little, and debates that raged more than two thousand years ago, are still valid topics for discussion today.

I should pause here and address the question of who I am and my qualifications for discussing these issues. I'm a financial advisor, with over 20 years of experience and fieldwork in personal finance. I frequently come across folks who are struggling to bring balance to their finances. Many of them have doctorate degrees, so I can't just talk to these highly educated people about simple math, or budgeting. Some of them are high-income business owners, so their income is not the problem either. So for over two decades now, my role in hundreds of people's lives has been to support their efforts toward reaching their life goals, and finding out how these efforts can be improved.

Also, as a first-generation immigrant to the US, I certainly went through a period of isolation, financial distress, and a culture shock myself. I'm sharing this part of my personal story to simply admit that, I know what many of you are going through. For a period, I was also left behind. Once I was able to build a life for myself with a business I could hang my hat on, I was able to entertain great hopes for the future, and I felt compelled to share what I learned over many years so that this information could benefit others as well. In other words, I've been a case study for my own research, and that's why I feel confident and comfortable sharing all that I've learned.

Here in the US, our socio-economic structure and institutions shifted and put the burden on the individual to plan and save for retirement, college education, health care expenses, and more. I suggest that in fact, it's the individual that needs help from all parts of society in order to reach financial stability. We have the resources needed to be able to do this and much more. It's our priorities that need some fine-tuning.

Most of us don't want to settle down in a monastery to find inner peace. It would be ideal to find a way to balance the requirements of modern life and be able to free ourselves from the grip of our egos at the same time. With some work, it can be done. Can the same methods that help attain a more stable and quieter mind also improve our ability to create a healthier financial balance sheet? I believe the answer is yes, and in essence, this is the idea behind IWT.

I also intend that this book will serve as an inspiration for younger generations, to help them have the courage to see things differently, and to encourage them to demand that a new set of cards be dealt which are constructive, balanced, and humane regarding financial opportunity and responsibility. If you don't want to become a member of the 75% who are ill-prepared for retirement, as a young person, it behooves you to wise up, not fall victim to consumerism, save enough for your future financial obligations, and consciously live a simple, balanced life.

I propose here practical ways to overcome the challenges I've identified, but I don't prefer that this book be an intellectual exercise only. I like to think of this body of knowledge and ideas as a work crafted to trigger a paradigm shift — to bring all of us to an era of creating healthier behavior patterns.

Your current actions are leading indicators of what is to happen to you in the future. Your current status or what has happened recently are the results of your past behaviors. So, you can choose to take constructive actions today, which will improve your future. In this, if you focus on progress, not on perfection, time becomes your friend and the process becomes fun. Otherwise, if you seek immediate results, your disappointment may also be immediate, and the journey would become a drag and would eventually prove to be unsustainable. *Your goal is to be on the path, and trust the path will lead the way.*

In the following chapter, we'll take a look at the cultural, socio-economic, and political structures we live in. Unless you leave where you grew up, and travel to a new environment with different customs and traditions, you won't fully appreciate the impact that culture has on your value system, opinions, choices, and the actions that you take. If we're overspending and not being frugal, it could well be that because of cultural factors affecting our mindset, or even that due to the spirit of the particular times we live in (Zeitgeist), factors may not be conducive for us to do so.

Before we point fingers at the individual, there are a lot of rocks to be turned. First on our list, is the fabric of American society.

2
The Fabric of American Society

If I could name one concept that emerges as the single most core American value, the above-all moral authority, and in the abstract, a concept as American as apple pie, it would be the idea of individualism. Merriam-Webster's online dictionary defines it to be: "a theory maintaining the political and economic independence of the individual and stressing individual initiative, action, and interests." The key point being the position that the individual holds in relation to other entities and institutions, such as the State, the Church, or the common good. According to Jean-Jacques Rousseau,[7] individualism rejects the notion that what's good for the masses inevitably translates into what's good for the individual — or that the collection of what's good for an individual will naturally contribute to what benefits the general interests of society as a whole. *So according to this view, the individual is absolved from any responsibility to contribute to the common good.*

This view has become a part of the overall character of American society and has had the power to set the tone for all factions involved with our social contract, including: politics, the structure of our institutions, the responsibilities of individuals and the authorities over them. As Adam Smith, (1723-1790), once famously said, "There is no free lunch." Newton's Third Law of Motion teaches us that "For every action, there is an equal and opposite reaction." In other words, positioning the individual at the highest point of the totem pole with regard to the social contract, comes with a price: insufficient systems to support each person's overall wellbeing, their financial health, and the lack of a safety net. These vital considerations are left for each person to address.

There are different takes on what individualism means, but what's consistently emphasized is: it's a social norm which American citizens appear to agree upon, which they live by, and that they are prepared to be judged by — especially when it comes to financial consequences of individual choices. This, I found to be the most relevant definition to serve IWT, especially the resulting self-reliance aspect, and how that reflects upon ideas related to personal finance.

To understand how and why the individual gained this elevated status in the US, we need to view what happened through the lens of historical context. And to do that we'll explore a comparative analysis of all the related -isms, as it proves to be helpful in answering the vexing question — that is, why do people overspend to their own financial demise?

2.1 — Individualism

The individual didn't always have this elevated position in society. First, a long and difficult battle for power had to be fought against the god-kings of the early agrarian societies. And again later, battles were fought against feudalism and organized religion, a process that went on until liberal democracies emerged in 18th c. Europe.

We can draw a nearly vertical line between the individualistic and collectivist cultures on a World map. Individualism is mainly a Western idea, while Eastern cultures, including Eastern Europe and Russia, carry more collectivist characteristics.

In Asia, this difference is even more pronounced. Particularly evident in the Eastern traditions Buddhism, Taoism, Hinduism, and Confucianism, the "self" holds a different meaning, which impacts their idea of individualism. Compared to their Western counterparts, the members of Asian cultures tend to be more aware of, and assign more value to their place in a group. As a result, they defer to, and are more prepared to conform to the needs of their society. In Muslim cultures, "ummah," the spiritual unity, is elevated to a stature above all else. The word "individualistic" has a pejorative meaning in daily language, and is used as a synonym to describe being selfish. As we will see in later sections, American, (and British), individualism differs from the "Romantic" version in Germany. Additionally, it's important to note that there's a direct link between the individualistic character of a society, and its level of democratization and economic mobility.

In the West, as part of the wave begun by the liberation movement during the 1789 French Revolution (with its Age of Enlightenment philosophies), individualism gained momentum in France in the 19th century. In a dialectic manner, it was a critical response to the Owenites, a 19th century utopian socialist movement led by Robert Owen. Owenites argued that individualism was nothing but a tendency to anarchy, which has no ideology other than what serves the egocentric needs of an individual at any given moment in time.

Steven M. Lukes[8] wrote:

> "In Germany, the ideas of individual uniqueness and self-realization — in sum, the Romantic notion of individuality — contributed to the cult of individual genius and were later transformed into an organic theory of national community. According to this view, state and society are not artificial constructs erected based on a social contract but are instead unique and self-sufficient cultural holes."

The evolution of individualism in Germany took a unique turn: proposing that the genius of the individual, although self-sufficient, is still part of a national community — an organic branch of a larger sum comprised of the State and the society.

Luke goes on to describe that in the United Kingdom, individualism allowed for religious nonconformity, which paved the way to its Church of England and its various versions of economic liberalism (including both laissez-faire and moderate state-interventionist approaches).

As previously mentioned, cultural differences have had an impact on the evolution of the concept of individualism and its forms. As we move toward more Anglo and less Saxon in the Anglo-Saxon lineage, as expected, economic implications and secularism gain strength, while the notion of Germanic Romanticism diminishes. The concept of an individual stops being an organic branch of a larger sum and morphs into an independent member of the group.

"In the United States," continues Luke, "individualism became part of the core American ideology by the 19th century, incorporating influences of New England Puritanism, Jeffersonianism, and the philosophy of natural rights.

American individualism was universalist and idealist, but acquired a harsher edge as it became infused with elements of Social Darwinism. *'Rugged individualism' — extolled by Herbert Hoover during his presidential campaign in 1928 — was associated with traditional American values such as personal freedom, capitalism, and limited government.* As James Bryce, the British ambassador to the United States (1907-1913), wrote in *The American Commonwealth* (1888), 'Individualism, the love of enterprise, and the pride in personal freedom have been deemed by Americans not only their choices but their peculiar and exclusive possession.'"

For the purpose of this book and IWT, this is where the rubber meets the road. The phrases Luke mentions above, "Social Darwinism" (referring to survival of the fittest), and "Rugged Individualism" both deserve our attention. Nowhere in the above ideologies do we see a mention of the common good, individual wellbeing or welfare, inclusive government, or the notion of a "national community," as we see in the German version of individualism.

And finally, again from Luke's article:

> "The French aristocratic political philosopher Alexis de Tocqueville (1805-59) described individualism in terms of a kind of *moderate selfishness* that disposed humans to be concerned with only their *small circle of family and friends*. Observing the workings of the American democratic tradition for Democracy in America (1835-1840), de Tocqueville wrote that by leading 'each citizen to *isolate himself* from his fellows and to draw apart with his family and friends,' individualism sapped the 'virtues of public life,' for which civic virtue and association were a suitable remedy.
>
> For the Swiss historian Jacob Burckhardt (1818-1897), individualism signified a *cult of privacy,* which, combined with the growth of self-assertion, had given 'impulse to the highest individual development,' which flowered in the European Renaissance.
>
> The French sociologist Émile Durkheim (1858-1917) identified that: the *utilitarian egoism* of the English sociologist and philosopher Herbert Spencer (1820-1903), who, according to Durkheim, reduced society to *'nothing more than a vast apparatus of production and exchange.'"*

To recap the concepts listed in the italicized phrases above, individualism, especially as seen in the US, is about: moderate selfishness, keeping to a small circle of family and friends, isolation, a cult of privacy, Social Darwinism, and a particular utilitarian egoism.

So, let's ask a critical question: can a person living with the above set individualistic values, still be happy and content?

In this environment, can the individual escape a crippling feeling of loneliness and be able to form strong relationships? Is it easy to balance being individualistic and still be an integral part of a social group? Would anyone be surprised by the level of widespread depression in the US after reading the above recap?

We are social beings, and a sense of loneliness is probably one of the harshest realities to emerge as a direct result of individualism. Happiness and being part of a community are intimately related. Loneliness and happiness, rarely co-exist. There's a reason why isolation in a prison cell is one of the harshest forms of punishment.

Responsibility and authority go hand in hand. The American individual has a higher level of freedom when it comes to life choices compared to their European counterparts, but this freedom carries with it a bigger responsibility, and that is self-reliance. *It appears that in the developed world, eased restraints upon what a person can and cannot do, come with a reduction of built-in safety nets to be provided as well.* This might have been an ideal compromise at the time of the drafting of the Declaration of Independence, but more than two hundred years later, this relationship deserves a review in order to meet the demands of today. The right balance between individual authority and responsibility, in order to bring the highest benefit to a person, should be an ongoing goal.

How is individualism connected to personal finance? The quality of safety nets in a society is directly linked to the acceptance and prevalence of individualism. The more individualistic a society is, the lower the bar on what will be provided by social programs (including pension and retirement funds). This is as direct as a connection can get, and that's why this financial piece is at the crux of any conversation about society and individualism. Some believe saving for future needs should be left up to the individuals. This could work except:

THE DECLARATION COMMITTEE.

1) In today's world, the extraordinary pressure to succumb to materialistic consumerism, is beyond most people's self-discipline and ability to resist. The proof is in the information shared in the first paragraph of Chapter 1. (Unless of course, one believes that 75% of Americans are irresponsible individuals.) We've built a society around consumerism, defined success as inextricable from great

material wealth, allowed corporations to use every trick in the book to oversell their goods and services, and now we hope that the individual will do the right thing and impose self-control as needed.

This is equivalent to hiring a magician for the birthday party to entertain the kids, and hoping they can see through his tricks. It is not realistic and it simply won't work because the magician knows exactly how to distract and cause his audience to lose focus. That's how his tricks work, and that's why he gets paid.

Similarly, we've allowed the cost of education and health care to sky-rocket, leaving it to the individual to borrow to cover these costs themselves (in the range of hundreds of thousands of dollars). We point the finger at them and put the blame of trillions of dollars of debt solely on their shoulders.

2) Even if a person is ready to do the right thing, work hard, live below their means, save and invest, there might not be enough jobs out there that pay a wage good enough to build such a life today. As you'll see on the chart on page 55, there hasn't been any meaningful wage growth for more than 50 years for 90% of the population. Soon automation and AI will not only make grocery store cashiers and bank tellers obsolete, but also doctors, lawyers, and architects. So who will be left to earn, plan and save for their future? For the majority of the population, this is all devolving into strictly a survival game.

This much can be said: laissez-faire policies and Social Darwinism are deeply rooted in the American public's value system and its institutions' DNA. This directly impacts the money flow, its distribution channels, rules, and regulations and ultimately the selection of winners and losers in the system. In 2008, banks who lent sub-prime mortgages were at least as irresponsible as the borrowers. In the end, banks were bailed out, their executives kept their bonuses, but the borrowers lost their homes.

If individualism is one of the root causes of the financial predicament most people are in today, is there an alternative?

The natural opposing position is collectivism, and when taken as a whole, it is not easy to consider it a better option. That's one reason why I've thought of writing this book, to instigate a wider discussion beyond these limited choices.

To side with absolute collectivism, one would need to present valid responses to the criticisms most often leveled against it, such as those raised by Ludwig von Mises[9] (1881-1973), a champion of individual freedoms. As he saw it, individualism would automatically be under threat by any challenge to it. He argued that favoring a collectivist position brings with it a risk of attributing independent existence to the collective, which could lead to ignoring the goals of the individual. These terms are arbitrary and we'll never be able to agree on the definition of the collective or the criteria used to form it.

Nietzsche[10] wrote:

> "Somewhere, there are still people and herds, but not where we live, my brothers, here we have States. State? What is that? Well then, open your ears to me, for now I shall speak to you about the death of people. The state is the coldest of all cold monsters. It tells lies and this lie crawls out of his mouth: 'I am the people.' That's a lie!"

Maybe we can start thinking a bit more creatively than Mises, and Nietzsche, and ask for ways that can harness the positives of both collectivism and individualism for the benefit of the individual. We're at a point in which individualism, in its selfish, Social-Darwinist version of the US, doesn't come close to accomplishing that.

The goals of the individual (thesis), and the collective (antithesis), don't have to be mutually exclusive. We must test these ideas to find their happy medium. That's what the influential German philosopher Hegel (1770-1831) taught us in his dialectic method. There lies a synthesis between the thesis and antithesis, and the uneven playing field created by the New World is forcing us to search for it.

Unless checked and balanced with democratic institutions, a State can turn into a cold monster, that is true. History is full of its examples, like Mao's China, Stalin's Soviet Union, and Hitler's Germany. But the State is built, organized, and governed by individuals. It isn't a separate entity ruled by aliens from another planet. So, it can also be structured to serve the will and wellbeing of the individual, as long as we find a way to shape it accordingly, and demand it to be so by democratic means. It is easier said than done, but referring to Hegel again, progress is messy and it is a never ending process.

According to a recent poll by Victims of Communism Memorial Foundation, 58% of millennials in the US would rather live in a socialist or communist nation. A recent issue of The Economist put this topic on its cover with a similar headline. You can argue the methodology or the validity of these polls till the cows come home, but it's obvious that collectivism is gaining ground. Do we think that's because of too much or too little individualism in our society?

Libertarians and conservatives are typically concerned about protecting individualism, in favor of smaller government. This is a commendable position seeing how big and authoritarian governments of Putin's Russia, Erdogan's Turkey, and Orban's Hungary has impacted individual liberties and universal human rights. But they must also be seeing the threat presented by the rise of identity politics and tribalization across the globe. These trends typically pave the way to polarized populism, hijacked by demagogue politicians, and unless stopped early on, in a relatively short amount of time, can lead to the emergence of oligarchic states as in the examples just cited. The irony here is, *individualism is losing ground to tribalism as a result of the very policies designed to protect it* because what may have worked during the Industrial Revolution era, is backfiring in this New World.

When the earliest texts in favor of individualism emerged in England by James Elishama Smith, William Maccall, or John Stuart Mill, these writings were a reaction against the harshest forms of socialist ideas, especially regarding property, or were an effort to elevate the individual above the concepts competing with it at the time, such as the Monarchy or the Church of England. These arguments were made during the Industrial Revolution, which brought a forceful push toward fundamental social and economic change.

A lot has happened since then. Humanity has gone through modernism and postmodernism, the fall of empires, the rise of nation-states, progressive social movements, two World Wars, the atomic bomb, the Moon landing, and most importantly the Digital Revolution. So now, we would do well to assess this arguably antiquated concept, as it currently exists in its most harsh, Darwinian, egotistical, and narcissistic form — from the standpoint of the damage it has wrought upon the ideals of quality of life and financial stability.

The counterintuitive aspect of individualism causes us to ask how it is that this ideology of elevating the validity of the needs and wants of the individual, can ultimately end up acting against the best interests of that same individual.

There are three causes of this effect:

1) When everybody puts themselves first, as opposed to keeping an eye on the wellbeing of the society as a whole, a person lacking compassion can end up disregarding any consideration such as other people's needs, animal rights, the environment, or common good, and instead, look for ways to serve their self-interest *only*. So the sum of these individual parts may create an unbalanced, unhealthy, violent, depressed, unhappy society with haves and have nots, (and more precisely, with a lot more have nots).

2) An egocentric personality may lead one to fall prey to commercial interests, marketing campaigns, pyramid schemes, gambling losses, get rich quick traps, and may also suppress the voice of reason at critical times when it's actually needed the most.

3) Probably the most important personal trait that leads to financial success is self-discipline. Unfortunately, a person focusing solely upon the immediate satisfaction of their individual needs will likely lack the required checks and balances, and healthy habits to stop consumption today, long enough to have the restraint to save for the future. *Egotism and the restraint needed to the right thing, rarely co-exist.*

There's a spoiled or injured child in each of us, whose natural inclination is to eat as much ice-cream or drink as much wine as we want, and do what ever it is that we want to do, unfettered by self-control. If you're not careful, that child/ego will spend all your money on unnecessary, feel-good, or luxury items, with no end in sight. Individualism encourages this inner child to scream ever more loudly, whereas collectivism aims to tame it. We need to get a handle on that inner voice, because otherwise one day, we could find ourselves at the doctor's office, diagnosed with diabetes, alcoholism or, buried in debt and being stuck with a medical bill that could force us into bankruptcy.

We'll discuss how this little kid with an insatiable appetite for self-destruction can fall prey to the corporate profit-making machine in the next section on consumerism. These two -isms, individualism and consumerism, amplify each other and put a person in a bind, leading them to unwittingly act against their own self-interest as they voluntarily empty their pockets without anyone holding a gun to their head. The synergistic effect between individualism and consumerism works so smoothly, that it's shocking once you're aware of it. If an individual doesn't learn to cultivate self-discipline and restraint, anyone with the right message can reach deep into that person's pockets, and do so on a daily basis. A few dollars here and there may not seem like a big problem, but add those dollars up and multiply them by decisions made across a span of 30 or 40 years, and you'd likely be shocked — and hopefully, you'd be motivated to stop overspending.

2.2 — Consumerism

If I wanted to summarize this book in one sentence I'd say, if you're having a hard time saving money, it's because you're a victim of your ego and consumerism. Consumerism is another child of the Industrial Revolution, and it's closely tied to individualism. In their modern-day versions, especially at the hands of an untamed ego, they can function effectively as obstacles to a person's wellbeing. Consumerism, like an amalgam filling, may have served people well in the past. But now, we live in a New World and we have to ask the right questions and make adjustments In line with today's realities. It seems ignorant, or downright destructive, to continue down this path, especially considering the toll on the environment.

In essence, two ideas are put forward here:

1) Individualism has been captured and held hostage by materialistic consumerism, and

2) Corporations, *not the individual*, have emerged as the benefactor.

The result is the triumph of the corporatocracy, which doesn't serve the individual financially, or emotionally, and is even detrimental to the earth environmentally.

Had the individuals been able to effectively use consumerism to demand free education, sustainable economic growth with environmental protections, equal opportunity, and affordable health care, then it would be harder to argue against it. Unfortunately, it isn't what we're seeing today. Consumerism, like other -isms, is a tool that is used differently in the hands of different societies with different priorities. We can revisit these ideas and use them to our benefit, as long as we're conscious of our consumption choices. With laws and regulations in place to support ideals such as environmental protections and consumer rights, one partner in these efforts will inevitably have to be — the State. It's up to each one of us individuals, to support politicians who will take up this agenda.

The principal assumption of free-market economics is that consumption decisions will be made by informed and rational consumers. It turns out, we're not as rational as we think we are. (More on this in the chapter dedicated to behavioral finance.) And secondly, corporations now know a lot more about our thought processes than we know about their products, or even our own consumption choices. Here, corporations have an unfair advantage in the race to our pockets. Through the use of opt-in 'cookies' which track our online interests, searches, and purchases, and with the information we so eagerly share on social media, every single aspect of our lives is out in the open for anyone to access. When it comes to the flow of information, there continues to be an enormous asymmetry favoring the interests of corporations and governments over those of the individual.

For instance, finance, for the most part, is a zero-sum game. Someone has to lose for someone else to win. In every financial transaction, there's a buyer and a seller, and by definition, one of these parties will ultimately be proven wrong. A professional investor's job is to be on the winning side more often than not. In the long run, the individual, or in this case the retail investor, is set up for failure, because they don't have the data and resources that are available to institutional investors.

In Western Europe, but more specifically in Britain, as the methods of production grew after the Industrial Revolution, so also grew the number of goods and services available to the middle and lower classes. Up to this point, except for a small group of elites and members of monarchies, people were poor, possessing only the essentials they needed: simple forms of food and shelter, a few pieces of clothing and personal items, tools of their trade, and essential furniture. From the end of the last Ice Age, this remained a twelve-thousand-year status quo.

So naturally, when the refrigerator came along, it made a huge impact on quality of life, and so did the oven, telephone, radio, combustible engine, light bulb, etc. What used to be a luxury only available to the very wealthy (things as simple as a mirror, hairbrush, or cookie jar), were now accessible to the masses. *So, at that point in history, the applications of individualism and consumerism did improve quality of life. In our current era, adjustments may be called for to restore the concept of quality of life.*

In *The Fable of the Bees* (1714), Bernard Mandeville[11] argued that vanity, high demand for more and more goods and services would be good for the economy, wages, economic growth, and ultimately the wealth of nations. Not only that, but he also addressed the Church's concerns by saying that rich people would become more honest, generous, strong, and honorable. He posited that poverty was to blame for creating the kind of weakness in people that diminished their integrity and pushed them toward immoral acts such as lying and stealing. He said that through the lure of fashion and advertising, people could be prompted to buy silly things they don't really need. He went on to describe that their resulting impulses to buy would stimulate economic activity, which would then empower governments to help the poor and those in need, (as well as help build necessary schools and hospitals for the society). As a result, the things that the Church wanted, consumerism would supply, while making people happier at the same time. So according to Mandeville, being rich and owning nice things could also be virtuous, while being poor was a form of living that rather hurt the soul. For the most part, Mandeville's arguments were popular and were widely accepted by many. Those were ideas from 300 years ago, when consumerism meant having access to most basic goods.

Jean Jacques Rousseau[7] (1712-1778), was one of the few who sided with the simple and virtuous model. He proposed that a good and meaningful life would look similar to the lifestyles seen in the small villages of Switzerland, and losing those would be a high of price to pay for change. He was a skeptic and held that concepts like improvement, growth, and development were overrated. He argued, that each step toward perceived betterment, was a step away from the simple and the pure, so he deemed it counterproductive in the end.

As it turned out, Mandeville's argument won by a large margin, and the proof is in the history that followed.

Back when I started my business in 2010, I was chatting with a friend, a professional in this field, about my firm's marketing strategy. He said to me *"Buğra, sell the sizzle, not the steak."* meaning that I should focus on selling the emotional benefits, not the features of my services. In the early stages of consumerism, the advertisement of goods and services was done precisely in the opposite way, that is to say, you'd make *the features* the talking points, not the benefits. The message to the consumer was mostly informative, because it was believed that in a free market, rational consumers would be able to make their own informed decisions. This presumption of the rationality of the consumer is a thing of the past.

Today, marketing is all about the sizzle, the emotional value registered by our lizard brain (more on this later). It's about happiness, as Don Draper, a character in the TV show *Mad Men* puts it, *"a call to our inner, subconscious desires."* Commercials urge us to make purchase decisions that all too often, are neither based on information, nor anywhere near serving us in our day to day needs. Rather, these offer emotional satisfaction, far from reason or rationale. So the answer to the question, why can't most people save money, is *that their subconscious desire received a call from Don Draper,* and it decided to give a green light to the next purchase. The rational mind will find a way to justify this in order to avoid any internal conflict (cognitive dissonance). We'll explore this concept more in the chapters dedicated to brain biology and behavioral finance.

We owe the emergence of the phenomenon of swaying public opinion through advertising to the nephew of Sigmund Freud, Edward Bernays[12] (1891-1995). He's referred to by many as the "father of public relations," and in the book titled *Propaganda,* he writes:

> *"The conscious, intelligent manipulation of the organized habits and opinions of the masses is an important element in a democratic society. Those who manipulate this unseen mechanism of society constitute an invisible government which is the true ruling power of our country.* (...) We are governed, our minds are molded, our tastes formed, our ideas suggested, largely by men we have never heard of. This is a logical result of the way in which our democratic society is organized. Vast numbers of human beings must cooperate in this manner if they are to live together as smoothly functioning society. (...) In almost every act of our daily lives, whether in the sphere of politics or business, in our social conduct or our ethical thinking, we are dominated by the relatively small number of persons (...) who understand the mental processes and social patterns of the masses. It is they who pull the wires which control the public mind."

Bernays was so successful at molding public opinion, or "pulling the wires which control the public mind," that his talents were recognized and utilized by the US Government of his era. President Wilson, in his attempt to convince the American public reasons that the US should enter the First

World War, stated that it was to *bring democracy* to Europe. Approximately 90 years later, the US invaded Iraq for the same stated reason. It seems like not a whole lot has changed over the years. In late 2019, as a result of a democratic process, the Iraqi parliament voted to have all foreign military leave their country. The US response was: don't push us unless you want sanctions forced upon you. Today, against the will of the Iraqi people, the US military still has a presence there.

Edward L. Bernays was so influential, that his methods and their effects could be the subject of a fine book on its own merits, and certainly deserves to be recognized here. For those interested in a lot more detail on this topic, I highly recommend the 2002 documentary by Adam Curtis called *The Century of the Self.* You can find it on YouTube with a quick search.

For those with no time to watch a four-hour documentary, the gist of the piece is that Bernays was the first person to promote goods and manipulate public opinion by using his uncle Sigmund Freud's[13] psychoanalysis findings. This approach, is based on the idea that human nature and behavior is driven by hidden, deep-rooted, and suppressed (sub-conscious) forces of sexual desire and aggression.

Freud had a pessimistic view of the individual. In his mind, if society as a whole were left uncontrolled, it could result in mass destruction. He thought humans shouldn't be allowed to express their freedoms, because at the root of it all was animalistic desires, and it would be too dangerous to leave people to their natural tendencies. *So, since democracy was the rule of public opinion, those in power should find a way to manipulate it and avoid this catastrophe.*

This is called *propaganda*, a word forever reviled as a result of Joseph Goebbels' applications of it in Nazi Germany. So,

Bernays thought, it should be renamed, and he came up with "Public Relations." Where propaganda efforts failed, Freud thought the populace should simply be controlled by any means necessary. He knew his would create discontent, but he felt sometimes it was needed for civilization to survive. In this, Freud clearly didn't believe in democracy, as he felt it reflected preferences of the poorly-guided masses.

Freud wasn't alone in his convictions. An Austrian politician, Adolf Hitler shared Freud's views, and he openly campaigned against democracy. In the backdrop of the Great Depression, the stock market crash, and social unrest, Bernays' strategy started to gain traction, and caught the attention of Goebbels, Nazi Germany's propaganda chief. This connection isn't a speculation, as Goebbels' own words reflected his admiration of Edward Bernays' work. The idea shared by these men, was *the need to control the masses and their desires, rather than be controlled by, and live in fear of, this presumed-destructive force.*

Around the same period, while Hitler was getting elected in Germany, the anger of the mob was more fortunately quelled by a wiser politician here in the US, Franklin Delano Roosevelt (FDR). It was definitely a remarkable time in history, where two industrialized nations and two political leaders, both facing similar economic troubles and social unrest, agreed upon the inefficiencies of the laissez-faire liberal policies, but went on to lead their nations down two completely different paths.

The New Deal in the US and the "public over private ownership" sentiment in Germany, supposedly had, in theory, the same goal: the good of the nation. The execution of these ideas strengthened democracy in the US and started a World War in Europe. Once again, the synchronicity can be seen in Goebbels' words of admiration for the New Deal as well. *The key difference between the two had to do with assumptions of public rationality or the lack thereof.* FDR believed in and trusted human rationality and the validity of his public's choices, while Hitler did not. The debate is far from over, even today.

Another long-standing debate continues to rage over governments' and corporations' ability (or lack thereof), to manage public expectations. FDR believed in engaging citizens via frequent polling. By doing so, the government could move in tandem with public opinion. The counter-argument to this, was to propose actions that would result in what government leaders wanted, having awakened the public's inner and hidden desires through propaganda. *Would the government follow the crowd or herd the crowd?* That was the question.

Consumerism then hit the stage in this argument, as an agent to help satisfy the inner desires of the individual. It was also expected to have the effect of making people more docile. At the time these ideas were being developed, another deep concern was the emergent problem of oversupply, resulting from improved industrial production. Consumerism shined there as well, creating conditions that contributed to a tide of excess goods. Once globalism began to make its presence known, the benefit that -ism promised to deliver was icing on the cake. There was and still is, always someone, somewhere in the world, who has an interest in purchasing leftover goods. So corporations kept on producing, without a care or consideration given to sustainability or damage to the environment of the planet.

Today, thanks to recent developments in brain biochemistry, we now know that consumerism's contribution to human happiness is a false promise, because happiness drawn from consumption is fleeting, and the slightest retreat creates a painful feeling of withdrawal. In that state, the docile citizenry has the potential to turn into an angry mob, and that may be the reason why the world's leaders are always obsessed with economic growth, with the goal of never taking even a thoughtful and needed a step back. Economists and politicians are so stuck in this gear, that feared economic shrinkage is oddly characterized as 'negative growth,' still growth — but negative.

So, consumerism was elevated to the level of cure-all status to address a perceived anger-of-the-masses, in contrast to Nazi Germany's public control methods. Over time consumerism has morphed into a religion of modernity. It's said that Bernays and others sincerely believed that the interests of corporations were no different than those of the American people. He contended that these manipulative techniques were used for the good of the people, (the same people, that he and like-minded others considered to be foolish and inherently violent).

A great example of how products are marketed and sold based on how well they satisfy the inner desires of consumers, even when those consumers are unaware, is an example from Betty Crocker. A certain cake mix in its product line failed to reach sales goals. A study was conducted, which revealed the reason was a guilt felt by housewives, at the prospect of serving cakes that were too easy to make. The mystery-barrier to sales turned out to be the factor of excess-convenience. By asking the housewife to whisk in an egg, sales of the mix soared. Participating just a bit more in the making of the cake made the target audience feel better about their effort and the end product. The subconscious desire instilled in women of that era, to be earnest, hard-working housewives was now satisfied by this mix.

Beyond the 1970's, the idea that people were inherently violent and ignorant, and should be controlled by all means necessary to ensure a homogeneous society, had only a few remaining adherents. By then, this stance was not only widely dismissed, but was even seen by many, to be part of the problem. The pendulum had swung to the other end of the spectrum, and a new sentiment created groups of people who tore at the social contract, or who ignored its existence altogether.

The essence of this new paradigm was that there was nothing wrong with the inner desires of the individual, and in fact, those desires should be set free and openly expressed. This was the "Zeitgeist" of the global movement beginning in about 1967. One fruit of this colorful movement can be seen in the 'flower children' of that period.

These two paradigms: one viewing the inner desires of the individual as dangerous, thus warranting outside control, and it's polar opposite, which felt there was nothing wrong with the desires of the individual (and on the contrary these should be freely expressed), did have two things in common:

1) In the end, they were both utilized by corporations to simply sell more goods and services.

2) They both fell short in helping to build a healthy and well-functioning society.

In the latter approach, an inflated sense of self emerged, free of restraints upon the satisfaction of inner desires. They were not to be suppressed, but instead were to be indulged in by people in power. This was the paradigm shift that occurred roughly after the 1970s. Consumerism would grow even more prevalent as a result.

As this new breed of non-conformist, free-living, and self-expressing individuals of the 1970s turned up their noses at the cookie-cutter products that had flourished in the past, corporations pivoted to conform to these non-conformists. By doing so, capitalism and consumerism triumphed once again, in the selling of even more goods and services. Individuals could enjoy an illusion of being free and anti-establishment, while the profits of corporations continued to skyrocket.

As a sidebar, I'll share a personal story here. From childhood all the way through college, I harbored a sense of resentment at my peers, as most of them seemed to me to be so intent upon following the latest fashion trends. Those who could afford to buy the genuine Western brands would proudly show off their fancy shoes, clothes, and watches. Those who couldn't would buy fake versions to keep up the ruse. These two groups usually didn't hang out together, as they belonged to different social groups/classes.

I simply didn't like what I saw, so I refused to be a member of either camp. My camp should reject both, and be free, I thought. As a result, I only owned a few essentials that fit into two suitcases (with which I moved to the US after college). I grew my hair long and felt very comfortable with the way I expressed myself as free and non-materialistic.

One day, a small group of my close friends got together and we went to see a rock concert. On the way back we took the city bus, and I was the first one to get off. As I stepped onto the curb, I turned around to wave to my friends, and what I saw shook me to my core: a bus full of college kids, who looked identical with their long hair and black leather jackets. The bus moved along, but I couldn't. I was shocked to view the reality in front of me: that while I felt I'd rejected "the system," and refused to wear its uniform, *I'd joined an alternative microcosm with its own uniform, which I'd freely adopted.*

This was a common everyday scene, but for some reason, the way it was so visually and starkly displayed in that moment, triggered an awakening in me. Everything I believed in, most importantly my identity, who I was, and what I stood for, was invalidated in my eyes. I wasn't as free, non-conforming, and rebellious as I had thought I was. I realized then that I was still conforming, though to a smaller group, and that was the only difference.

After college, I moved to California to earn an MBA, and in this vastly different culture, my whole value system and sense of self was challenged, broken into pieces and put back together, again and again, almost daily. Each time, I tried to put the healthier, more peaceful pieces back together, and leave the others behind. This is still an ongoing internal process for me — luckily, with less friction and suffering as time passes. There are some advantages of aging and gaining experience, thank God. In hind sight, I realize this was and is a process of unbecoming, more than of becoming.

But the main lesson I learned from this experience, is that *there is no such thing as "self," as most of us tend to understand it to be our "identity."* The self is nothing more than an idea, a mental construct, which can be altered, and reshaped. It's important to know that we can change for the better, evolve into a wiser version of ourselves, and become the person we want to be, rather than allow our circumstances to dictate to us who we think we should be. These circumstances include cultural biases and trends influenced by individualism, consumerism, and as we'll see later, materialism.

We'll return now to the beginning of the 1980s. During this time, in order to harness the individual's growing need for self-expression, large corporations and political parties in the West, especially in the US and the UK, began to organize scientific studies using focus groups. The goal was to identify and understand different personality types, and all the drivers that influenced each one's motivations. Ultimately, the gold in this information after being compiled and measured, had to do with how it could be made useful to sales. As a result of these efforts, it was revealed that the *individual's needs weren't all that unique after all.* These needs could be grouped into a relatively small number of categories — and most fascinatingly, the responses to the identified needs of these demographics turned out to be extremely predictable.

The results of these studies created the foundation of what would become known as "lifestyle marketing."

An interesting unintended consequence of this new type of individualism, including hippie movements, and the search for self-expression around the world, was a growth in support of right-wing, neo-liberal politicians like Reagan in the US, and Thatcher in the UK. They both got elected by promoting the individual above the government and by limiting regulatory restrictions on businesses. These two politicians received widespread support from socially-conscious, left-leaning voters, who at the time, found that they were aligned with the more traditionally conservative ideas of small government and relaxed corporate regulations.

These macro trends were perceived as a triumph of the individual over governments and corporations. "Don't shove your goods down our throats!" was the cry from the masses. The response from corporations was a pragmatic willingness to learn, adapt, and embrace self-expression. But once again, . saving and investing would take a back seat to corporate profits.

Consumerism has dominated social dynamics and politics since the early 1900s, and now even more so due to the influence of the digital age and populism. At first glance, this bending to the wishes of the public, may appear to be the basic functionality of a democratic, free and liberal society, but there are two problems with this. First of all, do we really have free will, or are we being manipulated and kept under pressure by social and cultural norms? Later in this book, Alan Watts will help us answer this question. Secondly, we're at a point now, where consumption-related motivations have become incompatible with most people's need and ability to build a nest egg. *Society's safety nets for normal people have been eroded, and consumerism-on-steroids has become the overriding social principle in America.* This has become a deadly combination that threatens the health of each person's financial net worth. On the one hand, people are able to find endless ways to express themselves freely through a wide array of products, but on the other hand, they've been shackled by an endless search for satisfaction through the spending of their hard-earned dollars, and even more so through borrowing.

2.3 — Materialism

Materialism is a concept with many forms that dates back to pre-Socratic philosophers. For the purpose of this book, I'll use the definition most pertinent to Economic Materialism. This is a personal attitude that prioritizes the acquisition of material goods. For most readers, it will seem so obvious to say that this doesn't bring happiness. But then, why do our actions contradict our thoughts so broadly on this matter? Why do most people behave as if the ownership of material goods will solve all of their problems?

Now, to be fair, it's common sense, and supported by evidence that the first $50 to $70 thousand dollars of a person's income in the US, does bring happiness. The simple reason is that this amount satisfies the first level of Maslow's Hierarchy of Needs: food, shelter, and clothing. Having these basic things makes all the difference in the world to your happiness, that's true. But as a person starts moving upward in this hierarchy, the law of diminishing returns kicks in, and the spending of every subsequent dollar begins to bring less and less perceived value. This effect is supported by evidence arrived at through research.

Somehow, it's easy to find ourselves chasing material success at the expense of everything else: our relationships, hobbies, helping others, and involving ourselves in meaningful activities. *We do this because most likely we've been brainwashed to believe that our identity and self-worth is defined by our worldly possessions, and how we are perceived by others.* It doesn't matter how many times you say that money doesn't bring happiness, the moment you identify a material object with your identity, it becomes shielded from all discerning radar, and is instead welcomed at whatever the cost.

Humans crave a "form" to hang on to, because without it we feel lost, untethered, unhinged, and vulnerable. Along with national, cultural, and religious identities, ownership of physical goods fulfills that need for a "form" or structure.

An interesting aspect of materialism, based upon what I've observed, is that those from upper-middle or high-income parts of society, don't rely as much on material goods to help them feel anchored or grounded, as do those from lower income levels. Wealthy individuals come to know that owning a house or land is a part of life, an expectation, for no other reason than as a natural progression of their eventual inheritance. So, young adults from affluent families can be perfectly happy with a minimalist lifestyle, spending their money on traveling and "experiences," rather than the accumulation of things. On the other hand, those who have never felt this type of safety, and can't even fathom owning a home, are more likely to waste money on cheap electronics, outfits, used cars, and fake jewelry, etc. *People who can't win big, tend to look for ways to compensate that loss by winning multiple small prizes at the expense of their overall financial health.* As a result, those born into affluent families with a built-in level of financial stability, get a jump start, and they are emotionally prepared to save and invest. While those born into less fortunate circumstances often live with a sense of financial insecurity, and this emotional state can trigger further self-destructive behaviors. The ultimate result is that the rich get richer, the poor remain poor, and unless mentally and emotionally this chain is broken, it's very difficult for the poor to navigate out of this syndrome.

Before the Industrial Revolution, people lived under the pretense of scarcity. Being frugal was a way of life. Today, the problem, especially in the US, is having easy access to too much "stuff." At the same time, the wage gap is growing and wealth is concentrated at the top. With the help of (social) media, people can now easily compare themselves to others' seemingly-perfect lives, and can feel shortchanged as a result. Judging by the number of anti-depressant prescriptions in use today, we can conclude that many people feel alienated, isolated, lonely, and depressed. This results in things being purchased for "feel good" reasons, such as to fill an emotional hole, or to relieve stress, rather than the simple functional purpose of the item itself. With the increase of the wage gap, borrowing has been made relatively easy, and as a result, debt levels have gone through the roof. Unfortunately, more and more ownership of goods doesn't address any of these issues. If anything, by depleting the very resources needed to have financial security, it makes things worse. This is the negative feedback loop many people find themselves in today.

In modern life, individuals have grown to be increasingly materialistic. Students are more likely to choose their field of study based on higher earning potential, as opposed to choosing a field they find personally interesting and stimulating. Businessmen and women go to work every day just to make money, rather than finding a profession that will give them the satisfaction of spending their time in meaningful activity while getting paid for it. Most companies have stopped producing high-quality goods and services which are intended to last a long time and build a noble legacy. Instead, they aim to sell as many products as possible, to as many consumers as possible, with the highest price tag as possible, at the lowest possible cost to make. A good example of this is fast fashion, with low-quality cheap clothing dominating the textile market of today.

So, are people happier, more content, and satisfied? On the contrary, *people who tend to focus on money, status, and "showing off," also tend to be less happy, and more irritable, lonely, and dissatisfied.* Materialistic pursuits fail to bring happiness, and instead can cause a deep disappointment.

In light of what we know about individualism, consumerism, materialism, and their combined effects on our pocketbooks, can we now conclude that these -isms are a big part of the overspending problem that many people struggle with? If so, what can be done about it? Let's discuss some potential solutions to these deep-rooted problems which have become so fundamental to Western socio-cultural and socio-economic algorithms.

1) Be aware.

As I've mentioned a few times, budgeting spreadsheets won't help most people. The race to access your wallet happens on a subconscious level, and the first thing to do is put out the fire at its source. Corporations have gotten much better at marketing to our lizard brain, partly because we willingly give them all the information they need to do so. If you want to get a hold of your life, especially your finances and purchase decisions, it is crucial to be conscious of the messages that enter your mind. Input and process determine the output. The only way to accomplish this is by being aware of what you watch and listen to, and be mindful of where your attention is.

A rough estimate is that 80% of every person's actions are driven by the subconscious mind. This is truly amazing to me. *That means we're basically bossed around without even being aware of it, and there is no conversation about how to alter this effect to work to our benefit.*

One way to shape your subconscious mind to be more in line with your needs is through the use of mindful awareness practices and affirmations. Every religion I've come across, attempts to achieve this goal by the use of prayer beads. With these, you can choose a selection of useful affirmations and create a ritual. As we'll see in the chapter on behavioral finance, rituals increase the strength of the placebo effect. With the help of repetition, you can work on re-programming your mind to serve your needs, and your actions will follow.

I was talking to a marketing executive friend some time ago and asked him to give me some tips on how to market my financial advisory business. He said, simply put, that I should focus on developing relationships instead. When I asked why, this was his answer: "A successful marketing campaign should have at least one of these two qualities; shock value, or repetition. Your profession doesn't allow for shock value, in fact it's practically illegal, so that's out the window. You don't have enough money for repetition either, so..." And he was right. Over the years I've learned that the way our subconscious mind can be shaped, is similarly either through shocks, repetition, or both.

We humans evolved as hunter-gatherers, and for both actions, attention is required. So if you want to control your actions, be mindful of where your attention is. *Your mind is the hand, your attention is the whip, and the action that follows is the tip of the whip.* If it helps, write this question down on a sticky note and place it where you'll see it frequently: "Where is my attention right now?"

I met a member of the US Marine Forces, Special Operations Command, and he had a tattoo on his hand that read: "focus." When I asked him about it, he said: "Losing focus can be deadly!" For us civilians, it can be costly.

It's critical to know *the triggers that push you in the direction of the shopping mall,* or to your next online purchase. Such triggers are all too often outside of the functional purpose of the object. Instead, the trigger can be the desire to chase away a sense of loneliness, boredom, depression, anxiety, or feelings of inadequacy or low self-confidence. Going to a mall becomes just something to do, and is a habitual social event for many. Once you identify your triggers, you can address them more proactively, and this awareness is the first step.

Let's say your trigger is boredom, then look for a better way to spend your time more meaningfully — like reaching out to a friend and spending quality time with them, *outside* the mall. Or go to a bookstore and pick up a magazine that encourages the adoption of a hobby that interests you. Perhaps sell some of your unused possessions for extra cash and to free up space. Along those lines, to stave off boredom you can make a conscious choice to focus on creating order: clean your house, organize your shelves, donate your books, and well, you get the idea. If it's a feeling of depression that's the trigger, you might want to take it more seriously and discuss it with a therapist, instead of looking for a band-aid solution at the mall. Whatever triggers your wasteful purchasing, the habit must be addressed. Otherwise it will always be there, and you'll never be able to start saving and investing.

The good and bad news is, that corporations are listening very carefully to everything you say and do. Literally, every time you make a purchase, you cast a vote with your dollars. But this means you can use consumerism to your advantage by voting for the right kinds of goods and services. Instead of a quick and easy fix, with a little bit of homework, you can align with companies that produce locally, support your schools, pay their employees a fair wage, and protect the environment.

Here's a precise formula: delay your purchases for a month or two, save and use cash for your next purchase, buy few but quality products each time, and make an effort to hold on to them for a long time.

Net Product Cost = cost of product x purchase frequency

When you lower the left side of the multiplier above, quality, it follows that there will be an increase the right side, frequency, meaning you'll need to buy it more often, and as a result, there is no saving to be had. Plus, buying this way contributes to creating more waste and depleting resources.

Speaking of using cash, if you want to get serious about curbing your spending, use cash only. Once your spending becomes tangible and you can see how much is leaving your pocket, it will be harder to let go of your hard-earned dollars. When you swipe a card, the effect is that it's just electronic numbers, until they are not and then the pain is real.

Also, there's nothing wrong with buying second-hand goods, especially when it comes to big-ticket items. Wherever you live, there'll be an affluent neighborhood somewhere close by. People with money own nice things. See what they sell online, and you'll be surprised the kind of deals you can find.

None of these are new or groundbreaking ideas. I just wanted to share a few reminders and pointers. For me personally, I can't hear a good idea enough. Repetition always helps me internalize points like these, and it may be true for you too.

A lesson from Buddhism: you can't fight darkness, you can only bring light to it, and it will disappear. Darkness is the absence of light, and your awareness is the light you want on parts of your life that can be improved up on.

2) Pay yourself first.

When you get paid or make a profit in your business, put aside a percentage, or a fixed dollar amount, before you spend a dime on other things, and deposit it in a savings vehicle. The best and easiest way to do this is by setting up a savings account for monthly transfers. Start small if need be, and you'll notice that you'll adjust to this saved amount's absence in your budget. Seeing that balance go up will in time become comforting. As this practice turns into a habit, you'll start wanting to save more, because it will morph into an action that makes you happy. You'll start releasing dopamine, a pleasure hormone, which will pave the way for your next deposit, because one thing we know about dopamine is that it's addictive (more on this later).

Shopping triggers dopamine production as well. So what we're doing here is replacing harmful habits with healthier ones. Once that shift occurs, you'll start noticing how much you've been overspending, which is exactly the awareness I was referring to a few paragraphs ago. This new level of awareness will in time rewire your subconscious mind, and you'll be less prone to make impulse purchases. Paying yourself first will leave less money in the bank for purchases, and unless you reach for your credit cards, which you absolutely shouldn't, it will force you to save. That's the whole idea.

When you acquire the self-discipline needed to save a small amount, the next step is to increase the amount and watch your balance grow even faster. If that puts a smile on your face then pat yourself on the shoulder, because it will mean you've been able to shift your status quo from 'unconscious spending' to 'mindful savings' mode. From then on, it's only a matter of time before your newly-acquired good habits take root. The next steps on this linear path to financial freedom are planning, investing, and monitoring, which we'll cover throughout the remainder of this book.

3) Have a plan.

Now that you're building a nest egg, you can start investing toward your life goals. Before that though, *you'll need a plan, as one has to begin with the goal in mind.* This will help you stick with your new processes and solidify your reasons to stay away from overspending.

Whether your goal is to invest for retirement, education, travel, or the purchase of a new home, you'll have a much better chance of reaching it, if it's written down. What gets measured, gets done.

To create an appropriate plan, I suggest that you work with a Certified Financial Planner,™ and in particular, one who's trained and who has earned the certification. If you can't afford to hire a professional planner for a comprehensive analysis, start with a small plan that your bank, broker, or even credit card company's website offers. A good resource for financial education and planning is www.cfp.net.

4) Invest.

Now that you have a plan, start investing accordingly. You shouldn't leave your money in cash for too long as it will lose its purchasing power against inflation. In Chapter 6, we'll discuss how to invest in a lot more detail. But for now, know that the US average long-term inflation rate is around 3%. At that rate, your cash will lose half of its purchasing power in 24 years. So, to avoid that, the least you can do is buy a 3-month CD and reinvest it every 3 months. That could be a great start for building emergency cash as well. A minimum personal goal should be reaching the Individual Retirement Account (IRA) maximum contribution limit, which is currently $6,000 ($7,000 if you're 50 or older). Not contributing to an IRA is leaving money on the table as your contribution is deductible from your taxable income.

If you're younger and starting out, a Roth IRA might be the best choice. The differences, pros, and cons between the two are beyond the scope of this book, but at the very least, you should know that your IRA contributions are deductible from your taxable income, growth is tax-deferred, and your distributions will be taxed at your income tax rate at retirement. Roth IRA contributions, on the other hand, are not deductible and therefore your contributions grow *tax-free* and won't be taxed at the time of distribution. Consider setting a goal to max-out contributions to one of these retirement accounts before making a major purchase in any given year.

5) Monitor progress.

As mentioned before, what gets measured gets done. Without measuring your progress, you can't know if you're on the right track. Are you going to be able to meet your goals? This question needs to be revisited at least once a year.

Our brains' natural tendency is to suppress painful thoughts and feelings, stuffing them down to our subconscious to bring us the temporary relief of delusion. So tracking our success in reaching our savings goals will help remove the subjectivity and bring down the wall of excuses we tend to hide behind. Tracking success is instrumental to raising our awareness, and reinforces our conscious effort to stick with our goals.

Our brain biology and behavioral protocols make it difficult to prioritize future gains over instant gratification and reward. In fact, as we'll see in later chapters, most of our decisions are emotional in nature, so the cause and effect relationships of our actions get overlooked in the process. A financial plan can help us understand the long-term effects of our purchases, help us quantify our goals, and help us limit our expenses.

Modern life in the US makes a priority of instant gratification via individualism, satisfies our inner desires via consumerism, and defines our identity via materialism. Consequently, it can become a daunting task to refrain from overconsumption. One philosopher's teachings can help us see the light at the end of the tunnel because he shows us how we can circumvent these roadblocks by finding happiness naturally and organically. That philosopher is Epicurus.

2.4 — Ideals of Epicurus

So far, we've looked at the reasons for, and the effects of, material consumption upon our finances, happiness, and well-being. After all, the point of life for most of us, is to be happy and live a good life. This topic was carefully studied 2000 years ago by Epicurus[15] in ancient Greece.

Born in 341 BC, Epicurus asked a simple question: what makes people happy? In his day, most philosophers were concerned with what was good and evil, or what was the nature of life and matter, so a school that studied the nature of happiness did attract a lot of attention. The school was even subjected to untrue gossip as a result, but in spite of the juicy stories told, Epicurus and his students took the task of studying happiness, and what gives humans pleasure, quite seriously.

Later in this book, is a whole chapter devoted to exploring the ideas of several prominent philosophers on the subject of how to build a good life. Epicurus is singled out here, as his school is the perfect antidote to individualism, consumerism, and materialism. Humanity has learned a lot from him with regard to addressing social malaise. Fundamental human needs are timeless and universal, so his teachings are useful for us today and are worthy of being revisited.

Epicurus was a humble and simple man, who lived a modest life. Unlike his fellow philosophers, he didn't drink wine but only water, ate very little, only bread and olives, and wore very simple clothes. He gave his life to understanding human wellbeing, and after years of study, he laid out three principles for achieving it:

1) Rather than spend all our efforts chasing a romantic or a sexual relationship, one should allow time for building meaningful, lasting, and satisfactory friendships. Epicurus observed how possessive, jealous, and mean-spirited individuals could be toward their so-called lovers, and yet turn around and be very understanding, supportive, and kind toward their close friends. This led him to believe that friendships helped people become better individuals.

   So he proposed to spend more consistent and quality time with friends, and to this end, he formed a commune where all of his friends and students lived together around the Epicurean principles. These communes later inspired Karl Marx and became the topic of his doctorate thesis on communism.

   Epicureanism in this original form, shouldn't be confused with the hedonistic, pleasure-seeking, selfish connotation it acquired later as the centuries stretched on. Back in ancient Greece, this -ism meant living a humble life, with basic necessities, in the company of good friends.

2) The kind of work that makes us happy isn't necessarily the one that pays the highest salary, but one that will impart the feeling that we're making a difference, helping others, and changing their lives for the better. The constant rat race of striving for more and more money, enduring demanding work schedules of grueling long hours, and other sacrifices in order to achieve a

successful career doesn't always translate to happiness. Instead, it can build resentment and take a physical and psychological toll on us as we're kept from doing the things that make us truly happy, such as spending time with good friends and close family, travel, and self-actualization. So in the process of choosing a career, we should look for these qualities, instead of putting all of the emphasis on the bottom line.

3) Luxury isn't worth pursuing. Big houses, fancy watches, fast cars, all make us feel good for a short period, then quickly become old news, and feel normal. Once the excitement of the new is over, the cycle of dissatisfaction returns, and the hunt for the next big thing resumes. In striking contrast to its intended goal of making us feel satisfied and happy, a luxurious lifestyle creates quite the opposite effect. What is luxurious today becomes rather dull tomorrow, and there is always someone living an even more luxurious lifestyle than our own. Rather than this shortsighted, never-ending chase, a simple life that takes care of basic needs and leaves room for leisure time (which can be filled with meditation, reading, taking nice walks, and ultimately room for reflection), makes us calmer, and happier.

The principles collected under Epicureanism were so successful, that the movement lasted for eight centuries. At its peak, it had about half a million people living in these communes, until which day they were taken down by the Christian church, and transformed into monasteries.

Today, under the light of neuroscience, brain imagery, and hormone analysis, we have enough information to confirm many of the findings of Epicurus. Humanity has evolved from a paradigm of scarcity, to abundance and wastefulness. We have tried consumption in search of happiness, but materialistic pursuits have failed to deliver. Instead, Epicurus advises us to *live simply, have good friends, find a career that feeds our soul and leaves us time for reflection.*

Today, as a whole we're physically more comfortable than we've ever been before, yet more people die from suicide than terrorism, crime, and wars combined. In most places around the globe, obesity is a bigger problem than famine. Our lives are full of luxury goods that the ancient kings and queens wouldn't have dared to imagine, and yet, despite all this, the rate of depression, drug use, psychological problems, loneliness, anxiety, anger, and discontent are on the rise. Something hasn't been working, and Epicurus warned us about all this 2,300 years ago. It turns out, not a whole lot has changed since then.

3

The US Political System and Its Drawbacks

This book explores why people overspend and fail to save enough for their future goals. It's impossible to present a case around this issue without looking at the political system that we live in. Whether a government incentivizes its citizens to save and invest or not, is probably one of the most important aspects influencing this problem. By the use of fiscal and monetary policies, governments control the flow of money. Tax policies alone can hugely impact the selection of winners and losers in an economic system. Labor and wage laws, the structure of social programs, and public retirement systems set the stage for individuals' income potential, saving, and spending habits. In fact, had the US offered a well-funded public pension system for all, I probably wouldn't be writing these pages, as there'd be no need to do so. As a result, a critique of the US political system as it relates to the purpose and main topic of this book, is included as part of the content of the Integrated Wealth Theory™ (IWT).

In the US, we live in a liberal democracy with capitalism and free markets. You may guess that all of my paying clients are actually the beneficiaries of this system, as they've been able to save and invest, enough so that they're able to hire me to manage their investments. So yes, I am well aware that it works well for some. But I do also see how individuals can and do fall behind. This obviously isn't a book on political science, and its scope with regard to that realm is limited.

Instead, this is an attempt to look at the consequences of the economic system we live in — which is a direct result of our political environment. My critique is on the conditions that have paved the way for the top *1% of the US population to own 75% of the wealth*. However, this is not a phenomenon unique to the US. Here is an even more striking fact; *the richest 62 people in the world own wealth equal to the bottom 3.6 billion people.* So in the big picture, it looks like capitalism works beautifully for those on top, and is supported by *those who are left behind, dreaming to get there one day.*

Following the fall of the USSR and the Berlin Wall, Francis Fukuyama[37] famously declared "the end of history," and the everlasting triumph of free-market capitalism. In hindsight, these words were premature at best. Global financial systems crashed 18 years later, and the rules of free-market capitalism were drastically altered. Today, even in the US, isolationism is being championed by President Donald Trump, and the trend is toward building walls and hostile international diplomacy. In the wake of the Covid-19 pandemic, even the most frugal countries such as Germany, have rolled out extensive aid and stimulus packages to help their economies stay afloat. In the US, the size of the stimulus is somewhere between 3-6 trillion dollars. It appears that the social state has come to the rescue of free-market capitalism, once again, twice in the last 11 years.

3.1 — Capitalism and Free Markets

We're at a turning point in history. Capitalism (which has brought wealth to many people in what we call the developed countries, and with the help of globalism, has brought billions out of extreme poverty in the developing world), is now failing to address the wage gap created by its very nature. This is an unsustainable trend, as in its later stages, it will leave no consumer left to keep the wheels of the system turning. What made economic sense in the past, appears to be vulnerable to shocks, as it frequently suffers through crisis after crisis.

One particular inner conflict is the progressively-suppressed wages issue. From an investor's point of view, profit margins improve with lowered costs, and employment is usually one of the largest items on the expense list. The means of production are privately owned and organized under a corporate structure. To increase market share and improve profit margins, companies compete for the highest possible earnings at the lowest possible cost. In most cases, suppressed wages or company-wide layoffs are seen as a positive development, which also push stock prices up.

When unemployment goes down, even though that's thought to be a positive for the overall health of an economy, it may put downward pressure on stock prices as higher employment numbers translate to higher costs and lower profits. *So, the very nature of capitalism is to pressure wages to increase shareholder value.* At some point, this starts to present its own problems, as low wages translate to low consumer purchasing power. This raises a critical question: *who will be left to buy the goods and services produced?*

In the US, 70% of economic activity is consumer-driven. When wages fail to keep up with inflation, this hurts consumer purchasing power and pushes demand downward. Lower consumption does eventually hurt the economy, *and* the corporations that have been lowering wages in the first place.

This problem is nothing new, and in fact, the concept was identified first by Karl Marx (1818-1883), (photo on the next page). Capitalism faced similar challenges in the past, and there's always been a solution readily available at hand. It's called outsourcing. The first time we reached this dilemma in

the 1970s, cheaper labor was available in the developing countries of Asia — mainly in China. Developing countries supplied the cheap labor needed, and the developed world had plenty of consumers who could afford to keep the demand up for goods and services. Global free-market capitalism pulled billions out of poverty, and the rich world gained access to a cost-effective means of production.

This win-win global trade system worked better than the competing alternatives such as communism, which fell behind, and ultimately collapsed in the late 1980s. With the help of outsourcing, the global economy was able to delay the problems that would arise from the wage gap issue. Fast forward to today, and there is even a cheaper labor source than that in the developing countries: robots and automation. So, no doubt a crisis is looming. Because automation not only replaces workers' labor, but also eliminates the consumption needed for profit-taking. It is a double-edged sword.

There's another large component causing the individual's plight today. Outsourcing wasn't capitalism's only solution for overcoming the wage gap issue. Starting in the 1980s, home, car, and student loans, along with credit cards, succeeded in creating a new demand that wasn't otherwise possible. *The "gap" was filled with consumer debt.* Today in the US, 80% of all home purchases are made with mortgage loans. This is encouraged by the Federal government's primary residence mortgage interest deduction and federal loan guarantees. Only 10% of the cars sold today in the US are paid for without credit. Of all the consumer debt, the largest balance is student loans, which currently totals above one trillion dollars. This uncontrolled, unregulated frenzy, along with the securitization of debt instruments, brought the system to a halt, resulting in the sub-prime mortgage crisis and stock market crash of 2008. It was said to be the worst global financial crisis since the Great Depression.

Here is what we've discovered so far:

1) Large corporations have access to almost unlimited production capacity, racing for the lowest cost possible, which pushes wages down as a result.

2) Central banks' money printing keeps interest rates low, credit plentiful, and banks become incentivized to take risks and keep on lending.

3) The consumer is under social pressure, and is the target of highly-effective marketing campaigns pushing them to increase consumption.

In addition to the above, in the following chapters, we'll see that our brains' evolutionary trajectory is conducive to making irrational decisions and overspending. Now, do you see why it's so extremely difficult to save money? There's nothing about this system that incentivizes one to do so, instead, the average person is pressured to borrow and spend. People are alone in this fight, and unfortunately, many realize this once it's too late, during what would have been their retirement years. For most people, their wages, and purchasing power haven't kept up with their expenses, especially housing, medical, and education costs. Many are being crushed under a huge load of debt and interest payments, and as the economy goes through booms and busts, so do people and their families. Even worse, while big banks and other companies that were deemed to be "too big to fail" were saved from a collapse in 2008, struggling individuals were left in the cold. People who are responsible for the worst economic crash since the Great Depression, weren't held accountable. In theory, the US is a free-market economy, and yet in practice, we've learned that profits are privatized but losses are socialized.

In the US, average real hourly incomes haven't improved a dime in the last half a century. Please look at the chart, below courtesy of the US Bureau of Labor Statistics, presented by Pew Research Center.[16] The green line is the real average wage growth, which has remained practically flat since 1964.

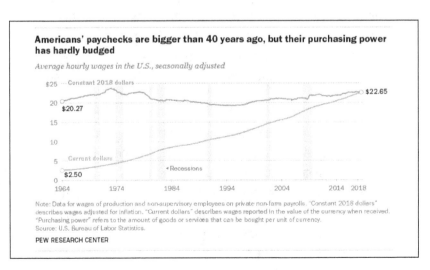

Americans' paychecks are bigger than 40 years ago, but their purchasing power has hardly budged

Average hourly wages in the U.S., seasonally adjusted

Note: Data for wages of production and non-supervisory employees on private non-farm payrolls. "Constant 2018 dollars" describes wages adjusted for inflation. "Current dollars" describes wages reported in the value of the currency when received. "Purchasing power" refers to the amount of goods or services that can be bought per unit of currency.
Source: U.S. Bureau of Labor Statistics.

PEW RESEARCH CENTER

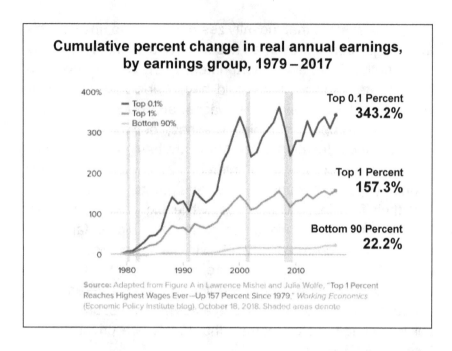

Cumulative percent change in real annual earnings, by earnings group, 1979 – 2017

- Top 0.1%
- Top 1%
- Bottom 90%

Top 0.1 Percent
343.2%

Top 1 Percent
157.3%

Bottom 90 Percent
22.2%

Source: Adapted from Figure A in Lawrence Mishel and Julia Wolfe, "Top 1 Percent Reaches Highest Wages Ever—Up 157 Percent Since 1979," *Working Economics* (Economic Policy Institute blog), October 18, 2018. Shaded areas denote

Now, let's have a look at what happened to the earnings of those at the top of the spectrum.[17]

Between 1980 and 2010, the top 0.1% percentile saw a 343.2% *real* growth in their earnings, the top 1% saw 157.3% and the bottom 90%, a whopping 22.2%.

In other words, for every 1 dollar of real growth that the bottom 90% saw, the top 0.1% gained 15 dollars. It's impossible to argue that the difference is due to those on top working 15 times more than a mine worker, or being 15 times smarter than a teacher.

The world is moving fast toward a structure similar to those portrayed in dystopian sci-fi movies. The wealthy-elite, 1% of the population, will be living in highly-secure gated communities, while everyone else will live in ghetto-like neighborhoods, subsidized by the government, and paid a universal income. The race to pay less and less for employees, and to optimize operational capacity, will eventually lead to replacement of humans with automation wherever possible. The likelihood that this scenario will play out, is surely only one of many possibilities, but without intervention it will come to pass with a high degree of probability. So now there's more than one reason to stop overspending. *Not only do you have an incentive to save for life goals you know about, you also have reason to save for unsavory unknowns which may be looming on the horizon.*

As usual, your biggest friend here is mindfulness. You need to wrench yourself out of blind adherence to political, religious, social, and cultural preconceived notions, and ask yourself this question: "How can I take the necessary steps to live a life with dignity, so that I'm not overwhelmed by changes coming in the next few decades?"

As is so often said, the only constant in life is change. It's impossible to predict the future accurately, but looking back at the last 200 years or so, we know that we'll have more technology in our lives, not less. So, we have to be flexible enough to adapt to it. In the spirit of putting financial stability front-of- mind, here are some pointers:

1) Know that in your job or business, the world only cares about what you produce right now, and at what price. In capitalism, what you've produced in the past is, for the most part, irrelevant. Accept that when someone or something can do your job equally or better at a lower cost, you'll be replaced. This doesn't mean avoid change, it just means you should accept the reality of change before it occurs.

   So, you have to be a constant learner and make an effort to keep up-to-date with whatever technological advancements will affect your career in the long term, so you can continue on your way to financial stability. To *stay gainfully employed, or have a profitable business, the key is to increase productivity.* In today's world, that means incorporating technology into your job before someone else does it for you, with or without your consent.

2) It's impossible to accurately predict future trends. Be ready for a change, even for a new career. The shifts mentioned here may unfold more slowly than anticipated. Look at the rate of online purchases, for instance. Today, less than 20% of purchases are made on the Internet. It was expected to surpass 50% by 2020. Also by 2020, we were supposed to have self-driving electric cars dominating the streets, which is still a moving target. More college degrees are becoming obsolete. Those with degrees having to do with computer science, robotics, bio-genetics, or who possess the technical skills necessary to design or operate a robot, install, uninstall, maintain, or do other useful handiwork that cannot be replaced by robots themselves, will have a seat at the table. In being open to learning how to use technology to improve productivity in your existing profession, you must prepare for possibly being forced out of your profession altogether. A true test of mental and spiritual youth, is flexibility and adaptability. We all need to find ways to stay young at heart, be resilient, and be open to letting go of what's not working.

Twenty years ago, it was said that paper would be a thing of the past, as all communication would go online. I met a gentleman who has a paper company, and I was curious about his thoughts on this. He said, "Oh, we can't meet the demand, there's not enough paper to go around!" I was admittedly surprised, until I heard the reason for the increased demand for paper: boxes. E-commerce is dependent on boxed goods and boxes are made of paper. So, most likely we will continue to see similar unintended consequences disrupting global trends, as opposed to easily foreseen changes, hence the importance I place upon flexibility and adaptability.

On the flip side, thanks to technology, services that were local and unscalable in the past (yoga teachers, voice coaches, accountants, doctors, etc.), are now able to take advantage of online communications technologies, and these occupations can now go global.

3) On the path of re-learning and re-tooling, it is important to adopt a longer-term approach. In other words, try to avoid the urge to save the day. *Instead do something today, that your future self will be thankful for.* This is an investment and like all good investments, should be seen from a longer-term vantage point. One thing I've learned from clients who were able to find great success and stability, is the importance of adopting a can-do attitude and not settling for the most easily-available option. Where there's a will, there's a way. In my experience, the right thing to do is often also the harder thing to do.

4) It may be harsh to say, but, keep in mind that capital-ism's priority is capital, not you the human, or you the individual. That's why *people die* today in wars, to gain *access to oil-rich* fields and lucrative trade routes. The lives of lost soldiers are less important to the capitalistic system, than the prospect of profits. That means, don't expect any sympathy from corporations or governments. The number one reason for personal bankruptcies, as mentioned before, is unpaid hospital bills. A compassionate mind has a hard time believing it, yet this is a reality and a risk for everyone in America. Until some sort of meaningful shift occurs to a system where *humans* are the priority, the only thing that can be done is to be realistic about it, and save for the rainy days.

5) There's an asset we all possess which is distributed evenly, and maybe that's why it's not given the value it deserves: this is time. In short, try to avoid wasting your most valuable commodity. This is the only thing in life, that once gone, cannot be replaced. In large part, you can often improve your health, you can re-make money that was lost, and you can fix broken relationships, but time is lost forever.

So, turn off your TV and limit your idle screen time. This might be hard to do with so many binge-worthy shows awaiting you, but I suggest that you set a limit to the number of episodes you watch at each session, or keep a book close at hand as a reminder to turn your TV or device off after one or two episodes.

You can't "unfriend" a relative, but it's important to limit the time you allocate to relationships that don't make good use of this precious commodity of time. We all have someone around, whose constant negativity drains us. I'm not talking about a healthy dose of skepticism here, which is a must-have for a critical mind. Instead, I'm referring to those who present you with undue friction every chance they get. Your time and energy is better spent doing more productive things, such as reading, resting, exercising, or other life-affirming pursuits.

I like the approach in this Buddhist adage "what you resist, persists," and this is true because it becomes solidified. Going around the roadblocks is usually an easier path. Like Einstein said, the best way to manage a problem is to stop it from becoming a problem in the first place. Similarly, the best way to manage time-wasters, is to avoid the situations most likely to lead to them.

I was talking to a business coach friend about a successful case he was working on. His client doubled productivity in

6 months. I asked him how, and the answer was simple: by helping him avoid distractions. Uninvited, unwanted distractions had cost this man half of his potential income, before he finally got serious about eradicating them.

Here *the key is being conscious* of what you're spending your time, focus, and energy on. If you have consciously allotted time to watch TV, that's one thing. If you got sucked into it and lost track of time, that's another. This is true of all your activities, and being aware of it adds up to conscious living.

Lastly, I'll share a question which I ask myself repeatedly, so it can serve to summarize what I want to convey here. It's a great litmus test that keeps me on track, and I hope it can help you too:

Is this the best use of my time right now?

3.2 — Abrahamic Religions and Dualism

"How we do one thing, is how we do everything."
(Zen Teaching)

What I put forward here, can generally be said for Christianity, Islam, and Judaism, or in short for the Abrahamic religions. This is due to the dualistic approach to God that these belief systems share, similarities in effects that each one has upon the human psyche, as well as similarities we see in the behavior of believers. Many social scientists see a direct link between a country's economic model and that country's prevailing religious beliefs. The most commonly known example of this relationship was introduced by one of the fathers of sociology, Max Weber, in his groundbreaking book, *The Protestant Ethic and the Spirit of Capitalism*[18]

(1905). He wrote that the Protestant work ethic was an important factor behind the emergence of capitalism in Northern Europe. He went on to write another two books on the economics and religions of China and India. He proposed that in the newer versions of Christianity, such as Calvinism and Lutheranism, luxurious spending was a sin, which helped with savings and investments. There's an undeniable correlation between the religious beliefs of an individual or society, and the corresponding model of their economic activity. We'll now investigate one particularly important aspect of this relationship — dualism.

In this context, dualism refers to the notion that there is a God, and everything else is created by It or Him. This commonality in Abrahamic religions places a hierarchic relationship between the believer, and that which is deemed to be the eternal truth. A distance emerges between the creator and the created, which contributes to the feelings of detachment, loneliness, anxiety, and depression. Materialistic consumption, addiction, violence, self-destructive behaviors, unhealthy relationships, and radicalism are some of the candidates ready and waiting to fill this vacuum of the heart. Unfortunately, a visit to the mall is usually at the top of the list of band-aid solutions, as that requires very little effort, and can yield immediate effects. This is this chapter's point, which is also relevant to the IWT.

I want to dive deeper into the idea of dualism, expand on its meaning in the context of this book, and put forward the cause and effect relationship.

As we all know, in Abrahamic religions, there is a God, a creator, the almighty power and source behind anything and everything that has ever existed and will exist for eternity. In this belief system, the believer is a member of the long list of created beings, along with all the animals, trees, rocks, rivers, planets, and on and on.

According to the Book of Genesis, God created man from the dust of the Earth, gave him the breath of life, and said "It is not good that the man should be alone; I will make him a helper, fit for him." And of course, as those writings go, it's then that woman is created.

The reason why I've lumped the Abrahamic religions into one is because this story is pretty similar in all three religions, and also in their different sects and branches. The creator is all-powerful, the eternal truth, all-seeing and knowing, while the *created are born with limitations as sinners, or at best, with potential to sin*, who then live predetermined destinies.

In Christianity, God is the Father, which is fitting because like a father, He rewards good deeds and punishes bad ones by sending us to heaven or hell after passing judgment, (even though He already knows the verdict at the outset). For believers, God knows which will end up in heaven or hell, and yet life, being set up as the testing ground, has to be lived regardless. But if the result of the test is already known, then what could be the point of it all? If God doesn't know outcomes in advance, is He limited in His powers?

A finding of psychoanalysis suggests that a mother's love is unconditional, whereas a father's affection and validation are contingent upon the child's performance in meeting family expectations. It may not be a surprise then, that the religions of history's matriarchal societies had no hell for the condemned, simply because there were none condemned in the eyes of the Mother God. As a result, the female deities didn't judge, threaten, or punish believers for their sins, and all souls go to heaven. Minoans of ancient Crete are a good example of this. For those interested in this topic, read the great book by Leonard Shlain[19] titled, *The Alphabet Versus the Goddess*. A local legend in Mill Valley, California, where he used to live, I had the pleasure of meeting him once after one of his talks.

Conversely, the male deities of the paternalistic societies, act more like father figures. This type of god keeps the believer under constant surveillance (bedroom and sexual orientation included). He collects lifetime data of good and bad deeds in order to make an informed judgment call as to which type of after life is most appropriate for each of his creations. There's no place to hide from feelings of guilt and insufficiency. *I speculate that one of the most common feelings of modernity, is not being or not having 'enough.'* Our brains are programmed to quickly look for ways to limit pain, and shopping provides the needed distraction for some, if not consciously, then subconsciously. This is in a nutshell, the implication of duality and its effects, which are relevant to the IWT.

Duality is everywhere we look. Always there are black and white, good and evil, back and front, old and new, happy and sad, fast and slow, peak and trough, friend and foe, open and closed, rich and poor. *We define and make sense of life on this planet, in a binary fashion.* There aren't very many words at all that help describe the state of neutrality. We are either energetic, or tired, and there's no in-between. The only way we define neutrality is by negative duality, but still in a binary fashion. We can be full, or hungry, and in-between is expressed by not being hungry. I feel neutral to black beans, I don't like them, but I don't hate them either, and being indifferent is perceived as passive and detached. Why can't we be actively neutral? Because thanks to duality, that's not how our minds have been conditioned, and for some odd reason, we never talk about this. When casually asked how we are, we never say, "I am neutral." Even when we say "I'm OK." it's not exactly like being neutral. This frequent reminder subconsciously takes the person to a mental binary test, many times a day. Am I lonely, or do I have friends? It has to be one or the other, nothing halfway. This is the binary bind that duality imposes on us.

Individualism and duality complement each other. In college, I got a part-time job at a large firm as a paid intern. One exciting activity, to break up otherwise mind-numbingly boring days of making photocopies and sending faxes, was the break that I'd take, running down to the cafeteria to enjoy a cheese panini. Back then I could eat as many cheese paninis as I wanted in a day without worrying about my weight — a luxury of the young. Each time I ran down to the cafeteria, I asked people around me if they wanted something back. One day, a supervisor told me: "You should quickly learn, that there are two types of people. You, and everybody else. If you don't, it is a tough life ahead young man." This is a textbook example of a dualistic, and individualistic way of looking at things. These two -isms, don't only co-exist, but they complement each other, like materialism and consumerism. There's no conflict between them. I'm hoping that different parts of this book and the IWT, have started to come together and make better sense to you, as the building blocks of a larger structure that they are. There's more work to be done though, and a lot more to discuss to complete the puzzle.

As an alternative to duality, we can say that white is defined by black, tall by short, back by front, life by death, big by small, and in this way they can't exist without each other. If everybody was tall, then nobody would be tall, but at the same height. We are connected to each other and our surroundings in so many ways. For instance, in traffic, our lives depend on thousands of other drivers paying attention to where they're going. Without their crucial focus, we could easily be in a car wreck, even if we did all that we could in our power to drive safely. We rely on the quality of the roads we drive on, the safety of the cars we operate, and the effectiveness of the traffic rules that bind us. None of these qualities are in our control. Every day in traffic, we jump onto a complex interconnected system, and the malfunction of one taillight could disrupt it entirely.

Our children's minds are shaped by teachers who, we may not ever meet. We take a pill prepared by a pharmacist we know nothing about. Our job performance depends on other participants' diligence. The pilot of the plane carrying our family becomes the most important person in our lives for the duration of that flight. I can go on, but it's fair to say that we're all dramatically connected. Once we see this, the idea of: us against them, me versus them, or black and white, dissipates. It's one big gray zone that we share with all other sentient beings on earth, through widespread interconnectedness.

For many, to give meaning to life, or answer the big questions of existence, life's purpose, and the direction we're headed, a spiritual guide or established discipline, is helpful. The anxiety created by not knowing what happens after life, the search for a universal moral code beyond cultural boundaries, the craving for there to be a higher power watching over us and our loved ones — and that one day all the injustice done to us will be made right — are all-powerful drives. For that reason, religion and spirituality have a high-powered position, and offer one of the most important protocols for decision-making.

If so, then we have to look under the hood, and see how these beliefs affect our money and finances. For instance, we have to be mindful of how much guilt is placed in our psyche by dualistic judgment. I am using the word "placed" as that's either knowingly, or unknowingly, with good intentions or not, what can happen. People who are raised under the influence of dualism and with a guilty conscience can be prone to developing self-destructive tendencies, and overspending is one form of this. Willfully giving money away, while being fully aware of its adverse consequences, fits this description "to a T." Some people fear success or having money, as they either don't think that they deserve it, or they fear losing it. Not having money can relieve the anxiety of a potential loss, so they may self-sabotage the prospects of this outcome. Gambling is a perfect example, and illustrates how a guilty conscious, caused by the effects of dualism, can lead to the doors swinging wide to financially destructive behaviors.

A common response to the paragraph above is, that it's not that simple. I don't claim it to be so. On the contrary, it's a complex problem and these are only some contributing factors. You can add yours to this list to compliment the IWT.

One direct outcome of dualism can also be the development of victim mentality. The sense of detachment and the lack of camaraderie can be so deafening, that everything that happens can be seen as an insult. Taking criticism in this state of mind would be akin to torture. Feeling angry and resentful when criticized, impedes better performance and growth. Collectively with other similar consequences of a guilty conscious, the effect compounded over a long period can leave a price tag of millions of dollars.

The point of this chapter and the information shared above is that money-related issues can be a sign of much deeper problems, and telling someone to begin to budget so that they can start saving, without having a conversation about these issues, will likely turn out to be a futile effort. That's why, in search of a path to financial stability, we have to ask bigger questions, such as:

- Why have I not been able to sufficiently improve my earning potential? Is it because I haven't been able to ask for a substantial raise?

- Do I truly believe that I deserve a promotion?

- Why can't I hold a steady job or a relationship? Is it because I cannot take criticism?

- Why do I spend beyond my budget? Is it because I seek quick fixes to my loneliness, depression and anxiety through materialistic consumption?

Money is a big idea, and solving issues around it need to employ similarly big ideas. Duality is also a big idea. It's an overbearing cultural component of our lives, and it is found everywhere that Abrahamic religions are practiced. *Whether you're a religious person or not is beside the point here.* These are unavoidable influences that penetrate our minds through the language we use, our attitudes and values, and so need to be recognized so that we can complete the puzzle.

Is there a spiritual path that isn't burdened by duality? The answer is yes. If you're interested in finding out more about this topic, learning about Eastern Traditions would provide a good start. To be more specific, Hinduism, (Zen) Buddhism, and Taoism, present non-dualistic views of the creator, and the created — as they are one and the same in these traditions. I can't emphasize enough the striking differences between the two algorithms, dualistic and non-dualistic, and how they shape our lives.

Non-dualism is exhibited however, in the mystic branches of the Abrahamic religions as well. Sufism in Islam, Gnosticism in Christianity, and Kabbalah in Judaism, are each heavily influenced by Hinduism and Buddhism. Alan Watts describes this notion of non-duality very elegantly, and you can easily find his work with a quick search online. I highly recommend his work on this subject.

In these non-dualistic traditions, the creator is in every form of existence. For the believer, this means there's neither a separate source to pray to, nor a separate source from which they'll receive judgment. The responsibility of being a better person falls solely on the individual. In Buddhism and Hinduism, if you fail to accomplish this, you'll simply come back in the next life and try again, and again, until you reach a state of freedom (Nirvana), from wheel of karma called Samsara.

According to this belief system, there's no self and others. Everything's an expression and reflection of the creator. It's all one source. So for the believer, non-duality by it's very nature doesn't allow for feelings of loneliness, or a vacuum to be filled by anything, let alone by material consumption.

This sense of oneness also renders morals and ethics into obsolete concepts, because for the faithful, all beings are one and the same. Hence the rules against harming your fellows, including lying, or deceiving them, become unnecessary, as hurting other people would mean hurting the self — the God.

Let's revisit the Zen teaching at the beginning of this chapter. "How you do one thing, is how you do everything." We can't control our finances without controlling our lives. The opposite is also true. How many people do you know who are in full control of their lives, but their finances are a mess? Not too many I bet. So, the answer to the question "how can we get our finances in shape?" is to start by getting our lives in shape. We can't get our lives in shape with small budgeting sheets, we need big ideas, and hence the pages allotted to explaining the effects of dualism.

The author, and the majority of readers of this book, will likely have been raised in the mode of duality. As mentioned, the effects of this are literally everywhere we look. So it proves wise to offer solutions for the potential sense of detachment, alienation, loneliness, and anxiety that can arise as a result.

Let's start with meditation, which is likely the most helpful self-improvement technique I've personally experienced. I don't know of any other activity that you can so easily add to your life, that while sitting in your own living room 20 minutes a day, (and paying zero dollars by the way), will give as big a benefit to your day to day experience of life. This is mindful awareness meditation. It's a scientifically proven method that helps you think more clearly, helps you be present, feel grounded, and have healthier thought processes. It lowers anxiety, increases mental capacity, and in general, helps you live a happier, healthier life. Sounds too good to be true? It surely did to me when I first heard about it — but as I've mentioned before, all of these claims have been quantified, studied, and proven by scientific research.

Specifically relevant to this chapter, meditation will help you feel more connected with everything and everyone around you, and will help you overcome feelings of detachment that may be due to a dualistic frame of reference. With an increased sense of unity, you may find yourself to be generally more content, and you may realize that you seek the satisfaction that material consumption provides, less and less. The frequency of your trips to the shopping mall will decrease because the frequency and intensity of the triggers that once sent you to seek relief there will also decrease. You'll find no shortage of books, online videos, articles, and smartphone applications on the topic of meditation, so I'll leave the details of this exploration to the discretion of the reader. If meditation sounds like a difficult task, listen to what Erich Fromm[37] has to say about it. In his book *To Have or To Be*, he recommends sitting comfortably and quietly in a chair, with as little movement as possible, with no external stimulation for 20 minutes a day, every day. Many experienced meditation practitioners use the words 'sitting practice' and 'meditation' interchangeably, because it's all about being able to sit quietly and be content. This is a simple and powerful exercise to help find inner peace, and it works its best wonders if done with disciplined regularity.

Yoga is another lifestyle change or hobby you might consider adopting. The original purpose of yoga was for the practitioner to break the boundaries of duality, and cultivate a sense of unity with life and with everything surrounding them. That's one reason why yoga and meditation go hand in hand. They are a natural complement to each other. The other reason that so many meditation practitioners also do yoga is because it does help to be flexed before you sit for meditation. You'll be able to be comfortable in a sitting position for a longer period of time, which will help you remain *focused*. Lastly, it's easier to calm the mind when the body is also calm after stretching exercises. In short, yoga goes perfectly with the theme of creating a sense of connection and unity in addressing the adverse effects of duality. It's also a great exercise for all ages as, when done properly, it's gentle on the joints.

Like everything else, yoga practice here in the US can be much more commercialized and competitive, and in many ways it differs from styles of practice in other parts of the world. Nevertheless, it's a great gateway to a healthier lifestyle. After practicing for a while, you'll begin to feel more at ease whether you're moving around or you're sitting and relaxing. More ease in the body translates to more ease in the mind, and to close the loop here, less distress that needs to be suppressed by an external stimulus like shopping.

If yoga isn't your cup of tea, it's not a problem. You can take a gentle walk, ideally in nature, around a lake, a creek, or anywhere you find a sense of tranquility and connection with nature. If you can make it a group activity, maybe as part of a sporting team or social club, even better. After spending time in such activities, you'll be surprised at the strength of relationships you'll enjoy with your team or club members.

Consider donating your time to a non-profit organization or a charity which operates in a field that's meaningful to you. In this, you can and should follow your heart, and doing so, you'll meet like-minded people, make friends, and will work with them toward a common cause. You'll have a good time, and you'll feel good about it. An unintended consequence of such involvement could include that you'll be more likely to meet successful people with philanthropic tendencies. These folks tend to be involved with nonprofits, and it's not uncommon to find that many sit on the boards of such organizations. It's said that your income is equal to an average of the incomes of the 5 people closest to you. I'll add that the traits you embody, are generally the same traits you'll find in the people around you. So you'll not only be more connected, but also better connected.

One big misunderstanding that materialistic capitalism has imposed upon us, is the idealization of receiving. We tend to think that the more we receive the better we feel, when in fact, it's the giving that makes us feel more connected and happier. Biochemical tests have proved this effectively, but the point can also be made mathematically, by what is called the Law of Diminishing Returns.

It's a principle of economics. When you have no money, and your uncle gives you $5,000, that will be one of the most meaningful gifts you'll ever receive. But if your uncle gives you $5,000 per week, after a few months, the proportionate value of each additional $5,000 will speedily become immaterial. After one year, each payment's additional value to your overall wealth will be less than 2%.

Here's the amazing part: the Law of Diminishing Returns doesn't apply when it comes to the satisfaction you get from giving. There are different hormones at play here. In receiving, similar to the habituation that occurs with drug and alcohol use, to get the same effect, each time you must receive bigger and better gifts. In giving, like a mother's love for her child, the more that is given, the more that is created. None of this is opinion, but rather is clinically proven fact. When you build trust with one person, it helps to build trust with another, as we'll see in the next chapter dedicated to brain biology.

Lastly, take a look at gardening. I have no personal experience with it, but people who are very close to me do, and I can see first hand its positive effects on them. Being involved with nature, having a physical relationship with it, and seeing the results of your efforts blooming in the soil, creates a profound sense of connection with life, and is calming for most people.

In the second chapter, we discussed how our cultural and political environments impact our choices around money. In this chapter, we explored how profoundly the spiritual views we hold can influence our shopping decisions. In chapter four, we'll investigate the biology of the human brain, and see how this information can help us better organize our lives, and therefore our finances. We're continuing to put the pieces of the puzzle together.

4

Evolution and the Function of the Brain

In this chapter, we'll discuss the biology and some functions of the human brain. Our goal is to uncover what roles they play with regard to our thoughts, feelings, emotions, and ultimately, how they may impact our behavior patterns. We'll look for ways that we can trigger pleasant experiences intentionally and naturally, and how to intercept and replace financially destructive behaviors, such as overspending.

Many of us overspend to our own demise, most often to compensate for a lack of satisfaction with other parts of our lives. Feelings and emotions are, for the most part, biochemical reactions, and the lab that produces these chemicals is our very own brain and nervous system. If we can learn how to trigger the favored chemicals through natural means and healthy habits, we may find we're also able to channel our feelings and emotions toward the desirable spectrum in a sustainable fashion. Understanding this mechanism well, and becoming good at its execution, can be a powerful tool in the war against self-destructive behaviors, including overspending.

So, let's review the chemicals at play, commonly known as the happiness hormones, and examine ways to naturally produce them, to help us avoid wasteful spending.

The word evolution usually denotes a period of progress and improvement, however this may not strictly be the case when it comes to our brains. Since the introduction of agriculture, which began after the last Ice Age (circa 10,000 BC), the size of the human brain has taken a step *backward*. The reasons for this are highly debatable. Some claim this was the result of a wheat-based diet, while others propose that we simply stopped needing to be smart enough to survive in order to pass on our genes. It was long thought that we humans improved our lineage continuously, such that each new generation would end up being bigger, stronger, healthier, and smarter than the previous. Though now it turns out this may not be true.

In his best-selling book *Homo Sapiens*, Nouriel Harari[20] demonstrates, how the wheat of agriculture domesticated humans, as opposed to the other way around. The "cavemen" were hunter-gatherers. Their diet was diverse, rich in vitamins, and easier to digest. Many experts claim that agriculture has contributed to the advent of illnesses that the caveman never endured, such as: dementia, diabetes, and cancer. In essence, the human brain is still evolving, but perhaps in the wrong direction. We want to get clarity on specifics that affect brain function, because this book is about decision making, and the brain is where our decisions are made. If we need a healthy brain to make healthy choices, then adopting a brain-friendly diet is paramount to our success. If we pair this with healthier behavior patterns in general, then we'll also have begun a clear and conscious path to a healthier financial life.

Before we get started, I feel the need to add a disclosure statement here. I'm not a medical doctor, psychologist, or neurologist, and I have the utmost respect for expertise, science, and scientific methodology. Recommendations in this part of the book are intended to serve as pointers and practical ideas, but please do check with experts before you follow them. If I were to repeat this disclaimer after each suggestion, it would be laborious for the reader, so please use discretion and ask for your doctor's advice before you act.

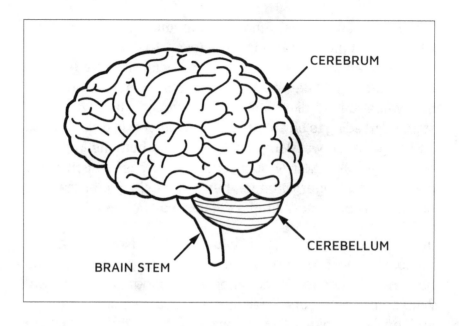

4.1 — The Story and the Parts of Our Brain

The human brain has 3 main parts: the brain stem (reptilian, lizard), cerebellum (limbic system, mammalian), and cerebrum (neocortex, the human brain).

The human brain has inherited structure from other primates, mammals, and even the first fish. So, while we can't isolate an exact timeline for the evolutionary progress of the human brain, we can say that brains developed in animals between 600 and 400 million years ago. We can also say the brain's mechanical development progressed in stages, akin to the layers of an onion. First came the nervous system and spinal cord, then the brain stem (continuous with the spinal cord), which, on crossing into the skull, connected the brain to the spine, then the mammalian brain, and finally, the neocortex. That's not to say other mammal brains don't have a neocortex, as in fact most of them do, but the complexity of the neocortex found in the human brain is unsurpassed in comparison to other animals. This is the structure that sets our brain apart.

4.2 — The Brain Stem

The brain stem is also referred to as the lizard brain, and it's responsible for regulating basic body functions, which we have in common with lizards — such as simply staying alive.

As the oldest part of the brain, it operates with no conscious thought required. Almost all of its activity is automatic, such as breathing, monitoring blood pressure and heart rate, eating and digesting, sleep cycle, vision, and other senses. Most basic body movements such as walking, standing upright, reflexes, and consciousness are also its responsibility. The brain stem is the bridge between the nerve connections of the brain and the rest of the body.

The brain stem has three sections. As the spinal cord enters the skull, it connects to the medulla, which is the first of the 3 structural parts of the brain stem. The second is called the pons, and the third is the mid-brain, (which resides right in the middle of the brain, hence the name).

We owe our vital body functions to the medulla, such as: breathing, hunger, heartbeat, coughing and sneezing, as well as vomiting. The pons is responsible for motor control, balance, and facial sensitivity, while the mid-brain controls basic body movements, the relay of information from the eyes and ears, and it wakes us up from sleeping.

If the second and third layers of the brain, (the limbic system and, neocortex, which together form the cortex), are seriously damaged, a person may go into a coma. But even in that dire state, if the brain stem is still functioning, the patient will be able to feel pain, and their body will have a regular heart rate and breathing. If, however, the brain stem has also been damaged and shows no activity, then the diagnosis is referred to in terms of the patient being 'brain dead.'

4.3 — The Limbic System

The limbic system, also called the mammal brain, brings the ability to feel and express emotions to our living experience — similar to other mammals, like our household pets: our cats, and dogs. These animals can make emotional contact as we do, so it might be easier to remember the functions of the limbic system by envisioning your cat's or dog's social skills. *Consider this allegoric definition, whereby, if your brain stem is your body, and your limbic system is your soul, then as we'll see in the next section, your neocortex is your mind.*

The limbic system, unlike the brain stem and the neocortex, is a network of separate parts, and is therefore called a system. It regulates our emotions and memories. Almost every emotion that makes us who we are, our social behavior patterns, relationships, likes, dislikes, sexual desires, religious beliefs and practices, and fight or flight responses, are all organized in the limbic system, which is why some call it our soul.

Here, we'll focus on the 3 parts of the limbic system, the amygdala, the hippocampus, and the hypothalamus. Of these, the amygdala is the part we repeatedly hear about as it's the all-important fear center of the brain. *Fears are powerful sources of our irrational behavior patterns.* Fear of not having enough in life, of losing what you already have, of not being loved, of being alone, or of not being validated, can't be isolated, but these are mostly attributed to the amygdala.

Marketing departments of corporations are way ahead of the game when it comes to using this piece of information on us. They employ experts in this field, whose job is to find ways to target the lizard brain and the limbic system in order to trigger emotions that will result in a purchasing decision. It's difficult to resist the temptation to overspend when a room full of experts has worked hard to convince you to do the opposite.

To situate yourself better financially, it'll help you to know how these tactics work, so you can overcome them by replacing your reactions with healthier habits to effectively stimulate your pleasure centers, instead of relying on consumption to accomplish this. Having a deeper understanding of the chemical cause and effect relationships affecting your brain, will help you make better decisions. Step by step, you'll build a more satisfying life that includes saving and investing for your future. If it's not for the goal of curbing wasteful spending, perhaps look at this information as a body of knowledge that will empower you to achieve greater happiness through healthier behaviors. For that end alone, it's worth paying significant attention to things that influence your level of consumerism.

The limbic system regulates complex emotions such as the fight or flight response, sexual desires, and hunger, in addition to anger and rage, and love and affection. It's also the center for happiness, motivation, and social behavior. Our limbic system is what helps us discern the difference between a happy face and a sad or angry one. It receives visual and auditory signals to recognize charming music or food that is good to eat. Without the amygdala, all of our social abilities would be diminished. For instance, it gives us the ability to recognize male from female, and an angry tone of voice from one that's friendly. I use the word "diminish" here, because there are other parts of the brain like the hippocampus that can regulate feelings of fear and anger as well, but the amygdala is recognized as the main center of these feelings.

The amygdala has two parts, one in each hemisphere of the brain. The fibrous tissue connecting its halves is larger in females. This is maybe why women are better than men at feeling and expressing their emotions.

In Ridley Scott's epic historical movie *Kingdom of Heaven* (2005), there's a pivotal scene right before Liam Neeson's character, Godfrey de Ibelin, succumbs to a fatal wound. He initiates his son Balian, played by Orlando Bloom, into knighthood with the following oath:

"Be without fear in the face of your enemies. Be brave and upright that God may love thee. Speak the truth always, even if it leads to your death. Safeguard the helpless and do no wrong. That is your oath."

The Baron then cuffs Balian with the back of his hand, and finishes the ritual "...and that's so you remember it!"

Why did Godfrey strike Balian? Is there a strong relationship between slapping someone in the face, and the strength of the recorded memory of the moment? And if there is, how did Godfrey know about it in the 12th century?

According to an article published in Nature Neuroscience[21] on Dec. 26, 2016 (see the link in the reference section):

"(...) Here we found that neutral stimuli encountered by human subjects 9-33 min after exposure to emotionally arousing stimuli had greater levels of recollection during delayed memory testing compared to those studied before emotional and after neutral stimulus exposure. Moreover, multiple measures of emotion-related brain activity showed evidence of reinstatement during subsequent periods of neutral stimulus encoding. (...) These results indicate that neural measures of an emotional experience can persist in time and bias how new, unrelated information is encoded and recollected."

In plain English, experiences that are attached to an emotion, get recorded in our memories more vividly. Without the presence of an emotion, the moment and the experience in it slips away from us and is forgotten. Maybe this is the reason why we're able to remember childhood memories which occurred decades ago, but what we had for lunch can slip our memory. As children, our emotions can be much more charged, so our memory of an event blended with those emotions becomes much more strongly encoded for recall.

The above phenomenon is strongly related to the amygdala's functions. Intuitively, and probably by the grace of common experience, people have always understood it. If you don't believe me, you can ask Balian. He never forgot his oath or the ritual during which he gave it. This was true not only because of the strong emotions unleashed by the initiation process, but also because he'd been struck by his dying father. Remember that song that reminded you of your first kiss? That emotional memory is stored very safely, courtesy of the amygdala.

In essence, our memory is an emotion-recording device, but it's far from being reliable. When we fail to take note of the facts of an event, it can be to avoid the discomfort they may cause us. Yet our memory will piece together some version of the reality in question, and it's likely to be colored by whatever emotions we experienced at the time. What's problematic, is that we naturally believe that our emotionally charged memory of an event, is how it actually did happen in reality.

We regulate our current moment with information pulled from past events, and plan for our future accordingly. *If this information is being recorded with the help of our emotions, then our future plans are also being made based on emotions.* This presents a huge problem that needs to be addressed. The solution? Neocortex-driven decision making, which we'll discuss later.

Next up in the limbic system, let's take a look at the functions of the hypothalamus. This is where, among many other critical hormones, oxytocin's production is initiated. Oxytocin is the "happiness hormone," so we have a specific section devoted to it. The hypothalamus is the bridge between the nervous system and the endocrine system, (the collection of glands that produce hormones and regulate such things as our metabolism). The hypothalamus, with the help of the pituitary gland, has the job of keeping our hormones in balance. This team controls our sleep cycles, growth, reproduction, lactation, thyroid glands, inflammation, and oxytocin.

An interesting aspect of the limbic system, is that it seems to embody what we identify as "self," as it regulates emotions and memory. If hypothetically you were to remove the cortex of a mouse, or a house pet, and then observe the animal, unless you're an expert able to identify the red flags, it would be hard to see a difference.

Why are we interested in this piece of information in a book about improving your personal finance? It's relevant because of the way our brain and its resulting priority algorithm has evolved. In our decision making, we're programmed to rely on the first two layers, the lizard and mammalian brains, the parts that evolved earlier, until we're forced to move up to the third layer — our rational brain. This process requires effort and awareness. *This algorithm is one of the root causes of the negative behavioral patterns which keep digging us into deeper and deeper trouble.* If you pay attention, you'll notice that most of your decisions are made wholly because of emotions, or at least under the strong influence of these two lesser-evolved parts of the brain. The neocortex, unfortunately, isn't accessed nearly as often as you'd think, or as you'd hope.

When important decisions are to be made, it's critical to find a way to move up to the third floor, and engage the neocortex. If you'd like to think of yourself as well-equipped to make the soundest decisions regarding money, then it behooves you to know about the ways the first two parts of your brain get involved, how they affect your behavior, and how to elevate your response to take advantage of the rational platform. The boss of that platform is the neocortex, and that's why we'll learn more about it next. It's the third and final layer in our onion of a brain.

4.4 — The Neocortex

The neocortex is the last layer to develop in the brain during its evolutionary process. The literal meaning of neocortex in Latin is "new bark" (neo meaning 'new,' and cortex meaning the 'bark on a tree.' As the neocortex surrounds the brain in a similar fashion, its name is quite fitting.)

Regarding personal finance, this is the most important part of your brain, because it's responsible for rational thought and planning, which are known as the executive functions.

The neocortex also regulates motor functions, such as the grabbing of a pen and writing, the steering of the wheel of a car, or shaving one's face. For the function of driving, the neocortex is also responsible for spatial awareness and visual processing. To make sense of sounds in traffic, as well as the very meaning of traffic signs, we rely on our neocortex. We also need it in order to talk to the police officer if we get pulled over, as it regulates the speech function — both for our speaking as well as for understanding others' speech patterns. Most neocortex functions help us with modern living.

One role of the neocortex which is particularly relevant for our purposes, and significant for everyday life, is decision making — a prime focus of this book. No one's financial success is assured. Even if you're born into a wealthy family, without good judgment, proper planning, and solid decisions, you can easily lose what you have. Conversely, doing the right things, like living below your means day in and day out, can add up to a decent amount of savings, and relatively quickly as well. With good investment decisions, your nest egg can compound to surprising amounts, which will be demonstrated in the chapter related to investing. All of this can be made possible if you let the neocortex be the driver of your life choices.

Specific to the neocortex, this part of the brain helps you control your more impulsive actions, and helps you make the tough calls when it's the harder thing to do. You can tell this structural area of your brain is engaged when you're able to resist eating that cookie if you're dieting, when you're able to make yourself choose the more economical car over the fancier one, and when you decide to move to a humbler home in a more affordable neighborhood. It's the last part of the brain to develop and it doesn't finalize its maturation process until the age of 25. This is why alcohol or drug consumption for youths under this age is much more dangerous than for adults. Cause and effect relationships aren't solidified yet, so a true sense of accountability is lacking until the age of 25.

A study published in 1970 by Walter Mischel and Ebbe B. Ebbesen at Stanford University, (now a book titled *The Marshmallow Test*),[22] studied children's decision making processes. A 4-year-old child is placed in a room with no distractions and is presented with two choices before the adult leaves the room. In the first choice, the child is told that a marshmallow placed on a table can be eaten after the adult leaves the room. However, they are also told that they have a second choice. If they patiently wait for the adult to return, and do not eat or touch the one marshmallow, they will be rewarded with a second marshmallow. Watching this test on video, and witnessing the internal and sometimes verbal communication that children have with themselves when alone in a room, is delightfully hilarious. It's also very revealing.

This conundrum provides a textbook test of frontal cortex function. At this early age, it's very difficult thing to choose to delay gratification, for the future promise of a reward for self-restraint. Yet, it turns out to be a very important ability, because what they found in this study, is that there's a correlation between the choices made in this exercise, and even future academic success (including SAT and college admission results), as well as eventual income potential.

This is the premise and the heart of this book, that we'll do a deeper dive into in the next section. *It's clear to me that many struggle financially because our cultural biases promote and reward instant gratification at the expense of financial stability.* If you don't have the money to buy that motorcycle, there is always credit available. After all, you're only going to live once.

That statement in and of itself is true in many aspects, but when you add that car, house, boat, fancy lunch, nice bottle of wine, expensive shirt, and other flashy products into the mix, you can easily find yourself lured into living beyond your means. It is easy to do. Our small mistakes when repeated frequently enough, can mount to financial disasters. The solution is in planning, restraint, and self-discipline, all of which are functions of the neocortex.

Then the question is, how can we bring the neocortex more into the forefront of decision making in our lives? How can we make sure it's the rational part of our brain that's in charge, as opposed to the emotional parts?

4.5 — The Triune Brain Model

So far, the way I've simplified and presented the human brain focusing on the function of its parts, is referred to as the Triune Brain Model, identified by Dr. Paul D. MacLean[23] in the 1950s. To recap, the three layers of the brain, in the order of its evolution are, the reptilian brain or the brain stem, the mammal brain or the limbic system, and the human brain or the neocortex.

As a summary: our reptilian brain, is responsible for the most fundamental survival needs, such as mating, regulating heartbeat, body temperature, and hunger. Its functions are automatic, unconscious and its main goal is to keep you alive. Aggression, dominance, worship, fear, and greed can be associated with our brain stem or reptilian brain.

In the early days of my investment career, one statement I heard over and over again from my senior advisors was, *"Investment is a game of greed and fear. People buy when greedy, and sell when fearful."*

Sometimes though, it's the fear of missing out that's driving a decision to buy, and greed which causes us to sell too early. When examined under this light, what a poor way to invest! If done in this fashion, some of the most important decisions of our lives will have been made by our reptilian brain! It's no wonder so many investors struggle with investing.

The limbic system regulates our emotions and memory. This is why we remember events that are emotionally charged. Our learning process is largely trial and error, and our emotional response to external stimulus plays a huge role in this. Since pain and pleasure are one filter we use to discern pleasant and desirable from unpleasant and undesirable emotions, our decision-making algorithm is largely like this:

- Seek activities that maximize pleasure and minimize pain.
- Do this with the least amount of energy and discomfort.
- Repeat as often as possible in the time allotted.

The limbic system rewards us for eating and having sex, and punishes us when we step out of our comfort zone and apply self-control. Shopping, or the gathering of goods for now and for rainy days, is extremely pleasurable, and difficult to refrain from. It's even earned the name 'Retail Therapy' in the *Urban Dictionary.*

The neocortex, the most recently developed part of our brain, is what makes us unique and human. It differentiates us from other animals and is the sophisticated center of logic, reason, and understanding of cause and effect relationships.

Awakening or enlightenment, as I see it anyway, is the recognition of this relationship. Living freely means being able to make rational choices that serve our wishes, without harming others.

The aberrant behavior of road rage, for instance, is exactly the opposite of this. It endangers us and other drivers, it solves nothing at all, it can't be explained by reason, and yet, anyone can fall into it during a weak moment of being controlled by their lizard brain. If you hear two voices in your head, that's because your lizard brain tells you to honk and yell, while your neocortex advises you to drive away. In this internal dialogue, the party that has been nurtured and given priority, will win the argument. It is up to us to decide which part of ourselves we'll let flourish, and which we'll subdue.

Freedom is the ability to ask oneself, "Is rolling down the window to flip off the driver behind me, the safe and rational thing to do?" It's easy to ask these questions before or after the incident, while we're still cool-headed, as our reptilian brain hasn't yet been triggered — and stress hormones aren't yet flowing in our veins. That's fair. But then it's also fair to say that in that instant, the driver exhibiting the road rage is being overpowered by their reptilian brain, and acting no differently than a scorpion fighting for territory.

Taking the body of knowledge presented here, and applying to your personal finances, I'm willing to bet that if you were to view your spending with an open mind, and ask yourself which part of your brain was driving this or that purchase, you'd find that, except for the very most essential ones, your reptilian brain and your limbic system, generated both survival and emotional impulses which provided the push to buy.

If you'd like some consolation, I'll acquiesce that you aren't solely responsible for this. With no exceptions, all of the goods you've purchased, or you're considering purchasing, have been advertised relentlessly to your reptilian and emotional brains, and not to your neocortex. Advertisers know that sex, fear, and greed are the most effective triggers for stirring up emotions and triggering the response they're after. Once your reptilian brain takes over, you're disabled from critical thinking, unless of course you *stop*, take a deep breath, and divert your attention to facts. It just takes some effort and practice.

This is where, that aforementioned marketing adage "Sell the sizzle, not the steak," begins to make sense. To market their products, advertisers use charismatic, sexy, and strong personas as their messengers, and the subliminal message (sizzle) is, that if you use their product, *you can be like them.* If you don't, you risk staying your boring self. And who wants that? Pardon the dramatization, but that's likely the internal dialogue going on deep in your subconscious, between your limbic system and lizard brain.

So, the advertisers taunt you: "Will you remain stuck in your drab status quo, or will you yield to our much more exciting suggestion? To reach your full potential you should go ahead and buy this shiny product, now! Call NOW! Even if you don't have the money, buy it NOW. We all know you have a credit card. If you don't, we will give you one, and charge you 22% interest. What are you waiting for?"

In fact, these experienced marketers are well aware that if you manage to wait until tomorrow, you'll have slipped out of your amygdala zone, and they'll have missed the window of opportunity to sell you that thing that you didn't need!

When these actions and choices are repeated, in time they can become habitual, and addictive because of the hormones secreted in the process (more on this in the next sections).

I'll share here a very useful technique I learned from my father, which has saved me a lot of headaches in life, and which has helped me leave the basement of my lizard brain and take the elevator to my neocortex before making important decisions.

My father was an officer of the Turkish military, and he told me that there, it's the law (or at least it was then), to allow a grace period before attempting to resolve a dispute among other officers or soldiers. When two people of uniform brought a complaint about each other, it would be timestamped, and revisited only after 24 hours passed. If the complaint did not stand after that period, it would be discarded, and no action would be taken. If however either complainant wanted to move it forward, only then would the escalation protocol initiate.

This rule allowed the time needed for a person to step out from under the amygdala's control, turn off the fight-or-flight response, reset with a full night's sleep, and realize that what happened a day earlier wasn't worth putting a written complaint in your fellow officer's file. In the end, there are two parts to every story, and it would be documented on the records of both parties anyway. After 24 hours, the *cost/benefit analysis* (one of my most-favored tools in the shed), would come to the rescue, and then more often than not, the complaint would be withdrawn and forgotten.

I can't tell you how many times I've found myself ready to say or do something in the heat of the moment, but instead chose to delay it for at least a day, and was so glad that I did. This 24-hour rule will not only save you a lot of trouble with your spouses, (I'm only guessing, of course, as I have no firsthand knowledge of this), co-workers, and superiors at work, but will also put money in your pocket as the frequency of your impulse purchases will decrease over time. Put a time-stamp on those important decisions, and revisit them 24 hours later, after a good night's sleep.

Arguably the best example of how we're driven to betray our best interests by our reptilian brain, is in the area of food consumption. We're still operating under the genetic coding of the Iceman, for whom food sources were often scarce, unpredictable, and geographically spread out. In the harsh conditions of his era, only those with the most energy-efficient physiology would survive, reproduce, and go on to participate in evolution. As a result, we can go without food for weeks as long as we have access to water, which can be plentiful in nature. Our bodies allow us to store protein, and have been programmed to run on stored fat extremely efficiently. We're able to respond to fight or flight situations at the drop of a hat by turning fat into blood sugar for the fueling of instant action.

Today, the equation is reversed, and our choices have been turned upside down. Food is plentiful for most of you reading this book, and in fact we live in a world where more people die from obesity than from hunger. We no longer have to chase a mammoth to feed our clan. All we need to do is to walk or drive to the closest grocery store, or put an order for prepared food to-go or for delivery. Without our reptilian brain nudging us, we'd only eat enough to satisfy our physical need for nutrients and energy, not more. But we all know that this is far from reality. Against all reason, our biggest health threat, diabetes, stems from wrong dietary choices. For many people, maybe as part of an oral fixation, the quickest answer to anxiety is the comfort provided by sugar, carbohydrates, or food consumption in general. When this is the case, the entire limbic system starts working against you, making that chocolate bar, bread, or pasta, nearly impossible to resist. Whatever influence your neocortex might have had in this situation gets brushed aside, and without the guidance of our neocortex, we're no different than any other mammal, just like our voracious household pets.

The ultimate question is then, who is in charge?

The results of scientific research suggest that the answer to the question above is, to a large extent, our emotions. That is, the automatic responses triggered by our limbic system and brain stem.

We've uncovered another piece of the puzzle here: the type of self-control needed for money-saving behavior, is driven by our neocortex, while instead, it's our limbic system and lizard brain that we generally allow to run the show. Regarding our brain functions, we certainly lose what we don't use, and what *is* used becomes reinforced. As a result, unless we're living a conscious life, we can easily fall into a negative feedback loop where our emotions become the drivers of our decisions — and with each new wrong decision reinforcing the next. Because of this, a big change needs to be undertaken: we must recognize when our limbic brain is being habitually engaged by a buying opportunity, and instead, in the heat of the moment, we must ask ourselves 'what would my neocortex do?' This will help us build a wider, and more quickly-accessed highway from our neocortex to our conscious mind. And this in turn will lead to a healthier feedback loop. Instead of the fleeting satisfaction of acquiring something new, you'll enjoy the satisfaction that comes with the success that is 'saving.' Just as with building muscle, regular exercise is the key, and the name of this exercise is: use mindful awareness (practices such as meditation), to interrupt the automatic selection of a particular self-destructive behavior. *This practice will help you build the desired pathways to your neocortex, and the resulting ride to your rational mind will get smoother and smoother.* Any activity that helps you bring your attention to your desired goal, is akin to taking your mind to the gym. The longer you can hold onto your focus without being distracted, the better shape you'll be in mentally.

So, how do we build the mental highways required to transfer those hard-earned dollars into our savings accounts? It can be easier than you think.

A simple mental trick to help you avoid impulse purchases is to employ a short checklist which you'll go over more than once before finalizing any purchase. It looks something like:

- Do I want or need this product? Wants and needs are not the same things. Wants can usually wait.

- If it is something I want, rather than I need, can I postpone it for a few months or a year?

- Can I function without it? If yes, why buy it?

- Do I have the funds saved or budgeted to buy it?

- Can I substitute it with a less expensive option?

- How often will I use it?

- If I won't use it frequently, can I borrow it?

- Is there a second-hand option available?

- Have I checked online providers for better deals?

- Are the customer reviews favorable?

- Is there a time of year I might be able to find it on sale?

- Am I on track with my savings goal for this year such that I can fully justify this purchase?

You can obviously add to this list, or edit it as you see fit, but do go through it each time you're looking to make a purchase decision. Using this process will help you disarm the automatic response of acting on emotions alone. By holding each purchase up to the light of this spending rationale, you'll be a witness to the steady development of healthier habits.

I'm sure you've heard of buyer's remorse, that sense of regret after purchasing something. I'm willing to bet that many of your possessions (which amount to thousands of dollars in spending), would not get a passing score if run through the checklist above. For many of those purchases, you've likely experienced buyer's remorse.

We can train ourselves to live a more aware and conscious life. Our neocortex has evolved specifically for uses such as these, so we're not completely helpless in this quest. We need to take the first step though. In the big picture, the goal is to separate feelings and emotions from rational thoughts, and then turn this into a habit. Here's a little game, or a challenge if you will. For the next 5 people that you have a conversation with, listen more carefully, and try to differentiate your thoughts from your emotions. In a relatively short amount of time, you'll find that most conversations revolve around feelings, hidden under *thought clothing*. Feelings are a natural and needed part of living a satisfying life. The trouble arises when we confuse them with our thoughts and allow ourselves to make decisions exclusively based on them.

In the following sections, I'll lay out how you can use your feelings and emotions, which are regulated by hormonal responses, to work *in your favor,* instead of against you. But for now, here's a simple technique you can use every day, and even multiple times a day, to strengthen your self-discipline: delay pleasurable activities, and prioritize the less desirable ones, and do so as many times as you can during a day. This is similar to going to the gym and exposing your muscles to tension in order to strengthen them. It's yet another mindful awareness practice similar to the ones we've already discussed, which will help you access your wiser neocortex.

Take this example: you know there's leftover cake in the fridge from last night's party, and you're ready for an afternoon bite. At the same time, dishes are waiting for your attention, or the lawn, or an uncomfortable email. Do the dishes first, and postpone the cake for an hour. Once you start looking for such opportunities, you'll find that there are plenty of chances to do this, and in time it will get easier to establish self-control. It's a physical and chemical process that you have control over, as long as you have the forethought and will to use it.

There are two points I'd like to make before we move on to the next section. The first is, now that you've been made aware of the different parts and functions of your brain, you can begin to ask the question, who's in charge of your conduct? For instance, the next time you see a TV commercial, try to see what thoughts, emotions, or feelings it ignites in you. Aim to test the triggers of your next purchases, and see what is under the hood — emotions, or reason? Needless to say, if you find yourself being driven by your emotions, please stop and delay the decision, revisit your checklist, and bring awareness to what's going on. The checklist strategy I shared with you is so simple that it's deceptively underrated. Test it and see for yourself. Print it and stick it to your fridge as a reminder, and you'll see that you'll be able to delay many of your purchases, and save thousands of dollars in the process.

Years ago, I heard what I've considered to be one of the most important questions of all time, especially as it pertains to business. While in a meeting a previous boss of mine asked: "Now that we've heard your feelings on this issue, what are your thoughts?" The colleague it was directed at looked quite embarrassed and uncomfortable, and I remember being glad I wasn't in his shoes right then. The point I'm making here is how important it is to separate our thoughts from our emotions. And I'm aware that as logical as it sounds, it's not always an easy task. When it comes to dealing with all the facets of personal finances though, it is a must-have skill to acquire.

4.6 — Biology of Happiness

What we define as our sense of self — through our perceptions, feelings, and emotions — is a collection of electrical signals running through our bodies, releasing specific hormones and neurotransmitters, which in turn trigger memories and behaviors. This is similar to the way that muscle memory works. That's why the simple act of delaying pleasurable activities and moving less desirable ones up on our list, can be so effective. Doing this breaks down and rewires our mental muscle memory. By doing this we create a gateway for becoming the person we want to be. We can consistently act in the way that serves our best interests, only if we take a moment to consider the sources of what's challenging us into action, and if we understand how to manage our reactions to these triggers. I'll likely repeat this sentence again and again, as it deserves our attention: *there's no such thing as the self. It's nothing more than an idea we've fixated upon.* It's a construct built of our perceptions and memorized reactions, all of which can be edited. If it can be altered, is there still a self?

We're a biological machine that functions at its best when in chemical balance. Once that balance is interrupted, we'll try everything in our power to get back into balance, even though it may hurt us in the long run, and may keep us in a negative feedback loop in trying to do so. Next, we'll look at how making use of some simple techniques will help us move toward that desired sense of balance, without our spending a dime on destructive feel-good purchases at the shopping mall.

Four of the most important neurotransmitters that affect our feelings and emotions are serotonin, oxytocin, dopamine, and endorphins. Most of our self-destructive spending habits stem from chasing one or more of these "happiness hormones." So *if we were able to identify and adopt ways to secrete these powerful chemicals in ways other than shopping, the urge to buy the next shiny object would become easier to overcome.*

Recreational shopping, unfortunately, is tough to steer away from because it's known to trigger not one but a combination of these addictive chemicals. This sense of satisfaction is always short-lived as we're highly adaptable, and we know all too well that what's new and shiny today, becomes normalized and mundane in our experience tomorrow. We easily become numb to stimuli. Consider how quickly we stop being aware that we have a watch on our wrist. *This keeps us sane by helping us avoid being overstimulated. Consequently though, we can find that we've become addicted to overconsumption, as a result of unconsciously chasing the chemicals which create a repeatable euphoria again and again, until we're officially bankrupt. There is a name for this: "shopaholic."* Just like there are varying degrees of alcoholism, one can become a functioning shopaholic in the search of emotional satisfaction through the constant acquisition of things. This affliction can remain hidden and unnoticed for years, until the hole it digs becomes too large to ignore.

4.6.1 — Serotonin

As the neurotransmitter known as the molecule of happiness, serotonin also has a part in the regulation of important things such as sleep cycle, mood, appetite, and memory. Low serotonin levels are associated with depression, social anxiety, and low self-esteem, therefore many modern anti-depressants are able to attribute their efficaciousness to their ability to improve serotonin levels in the brain.

If you're feeling down, or in a bad mood, it could be the result of a drop in serotonin. And it's interesting to note that this relationship works both ways, meaning that being in a good mood may actually cause an increase of serotonin levels. Another interesting thing about this vital neurotransmitter, is that 90% of it is produced in your gut, and as a result, your gut health has a lot to do with the mood you're in. Another quality of serotonin is its ability to improve *impulse control,* so it's production is extremely important to the focal point of this book. In short, arming ourselves with the knowledge of how to naturally produce serotonin would be extremely beneficial to the day-to-day quality of our lives, in addition to helping us stop overspending.

Here are a few things you can do, but remember to please check with your doctor first:

- Spend more time in the sun, around 20 minutes a day will help raise your D vitamin levels, which may improve serotonin levels as well.

- Exercise regularly. I'll beat this drum again and again, along with the recommendation to pursue a healthy diet, because it's so easy to go out and jog, or take a brisk walk. Choose an activity you can stick with to raise your heartbeat so you can reap its many benefits.

- Introduce food to your diet that helps with serotonin production such as salmon, eggs, spinach, bananas, seeds, and nuts.

- Try to avoid or limit gluten and eat organic as much as possible.

- Choose carbohydrates that have a low glycemic index such as brown rice and whole wheat-bread.

- Look for ways to improve your sleep, including going to bed before midnight, and making sure that you get 7 to 8 hours of quality sleep every night. It's important, and an entire chapter could easily be devoted to this subject. A little research will turn up many great sources of information on sleeping better.

- Have some green tea, turmeric, and dark chocolate, all of these are known to help with serotonin production.

- And the mother of all good ideas: drink plenty of water.

It's OK to fake it until you make it. Take a look at a photo that brings back pleasant and happy memories. Remember, memories are full of emotions. Positive memories will lead to positive emotions, and this will help you generate more serotonin naturally. Due to the way serotonin works, the production of positive emotions will encourage an upward spiral of positive benefits. That's good news. The bad news is that the opposite is also true. Negative emotions can inhibit serotonin levels, leading to a downward spiral into depression and other detrimental health effects.

Socialize with a friend you enjoy spending time with. Choose to recall memories of happy times. There's no shame or problem with inviting happiness into your current reality, as doing so will trigger beneficial brain chemistry in your present moment, and what may have felt forced 5 minutes ago, can become your reality in the now.

For the purposes of this book, serotonin's most advantageous function is how it effectively works in our brain as an agent of self-restraint. I haven't seen research that specifically links good serotonin levels with minimized addiction to shopping, but this neurotransmitter's ability to enhance self-control in the face of many forms of overindulgence is a foregone conclusion. It's useful to keep in mind that, in order to consistently curb unhealthy urges to spend, live within your means, make a financial plan, and stick to a budget. You'll do best to be proactive about making sure your brain has the levels of serotonin it needs, and in fact, healthy levels of serotonin may actually be *required* for financial success.

4.6.2 — Dopamine

Dopamine is a neurotransmitter associated with feelings of reward and pleasure. If there's one chemical in our bodies that can be directly linked to shopping addiction, this is it. It's interesting to note that dopamine is released during cocaine use, and that's the primary reason for the strength of its addictive nature. It boosts dopamine levels so high that without it, nothing feels pleasurable afterward. Our body's normally advantageous adaptability once again becomes a detriment in the case of this drug. If we can agree that an addiction to cocaine is bad for a person, so too is excessive shopping, and for similar reasons.

Some symptoms of dopamine deficiency include: lack of motivation, low energy, increased susceptibility to addiction, general mood disorders, and of course, depression. Aren't these very similar to what we see in serotonin deficiency? In this regard, yes. A negative feedback loop to watch out for here is stress and its effects. Stress hormones (such as cortisol), suppress dopamine, and in turn one may feel fatigued and moody. This may lead to recurring episodes of stress, such that steering out of this self-propagating cycle can be extremely difficult, especially if one also feels tired and moody. This is one reason why the initial intervention for depression is often chemical in nature, in order to jump-start the patient's brain chemistry to move up and out of this debilitating negative feedback loop.

When it comes to dopamine production resulting from the purchase of the next shiny object, the effect one gets from online shopping is amplified. The shopper is rewarded with good feelings while browsing the Internet, when the buy button is pressed, and it continues on through an *anticipation period*. This experience is re-lived again when the purchase arrives and the package is opened. In fact, often *dopamine is at its highest levels during the anticipation period*. This is one physical and biological reason why compulsive shopping can be addictive and therefore financially destructive. Studies show that a similar phenomenon is seen in other forms of addiction, like infidelity for instance. For many, anticipating the next lovemaking episode is where the addiction resides, not in the physical act. Even though afterward, the addict may experience regret, soon the urge to repeat the brain chemistry of chasing the next partner or encounter can overpower reason.

In the last 2 decades, the rise in US consumer spending has steadily exceeded the rates of inflation and population growth. This has resulted in, among other things, an increase in the average size of single-family homes and an expansion of success for the self-storage industry. Corporations that help us feel good about donating our unused goods never tell us, that most of these items end up in the landfill. Less than 10% of discarded plastics are being recycled. The environmental effects of overconsumption should be addressed in another book, but *fixing your finances by curbing excess shopping is not only good for you, but also good for the environment.* I can't help but see that for most readers, a whole year sworn off of frivolous shopping would be a very useful and direct way to pay down debts, improve balance sheets, and increase savings. And they may even find that they don't even miss it.

I've seen classes, workshops, and coaching programs that teach how to *trigger dopamine production* in people, with sales techniques, to put aside objections and lure them into buying products. The next time you suspect you've bought something you didn't really need, consider that you were probably the victim of one of these marketing tactics. Maybe keeping that in mind will give you the extra motivation to 'just say no" to those marketers and their tactics.

Since this neurotransmitter is linked to why we have a hard time resisting the urge to shop, let's take a look at some healthy ways to improve our dopamine levels.

If you'd count yourself among those susceptible to heartburn, gas, or bloating, these symptoms indicate an imbalance of stomach acids, which is another symptom of lower levels of dopamine. A weak digestion system could fail to absorb essential vitamins and minerals properly, which help with our health in general, but also help the production of neurotransmitters like dopamine. Both dopamine and serotonin levels depend heavily on gut health, so see your doctor if you continue to experience these symptoms. In the meantime, consider adding probiotics to your diet which may improve the levels of healthy bacteria in your gut.

As we saw with serotonin, stress counteracts the effects of dopamine as well. This causes more stress, and the downward spiral begins. So, we all have to find ways to lower or manage stress. I'll leave it to each reader's discretion here to choose the best method, as everyone is different. If supplements are needed, consider vitamin B1. There are also four easy-to-implement, free-of-charge, and life-changing habits that will improve your wellbeing so significantly, that combined, they can make up for a long list of harmful errors. Avoiding sugar and keeping your blood sugar levels low is one of them, and meditating, exercising, and socializing are the other three. They each have multi-faceted health benefits and can stand as your short list of ways to support optimum brain chemistry. You might also consider adding nutritional yeast to your diet (after talking to your doctor), as it can help your body produce a substance called uridine, which can keep your dopamine levels healthy, and also can help you get better sleep at night.

Taking in 20 minutes of sun per day has benefits by increasing your Vitamin D3 levels. By now, you can probably see a trend here. By implementing good habits for optimizing brain chemistry, you can improve your overall health, quality of life, and ultimately, also your financial wellbeing.

Reduce or eliminate altogether your alcohol and recreational drug use. Alcohol is a toxic depressant, and especially if your body lacks dopamine, it will make your situation worse.

Try to reduce your caffeine intake. It has many health benefits, but after your second cup per day, it could work against you by inhibiting your natural dopamine production. The good news is that decaf coffee has many of the same nutritional benefits without the caffeine overload. That being said, a cup or two a day may actually help trigger dopamine. But as with so many things, moderation is the key.

Eat more fruits and nuts as they're full of amino acids, which aid in the production of neurotransmitters like dopamine. A protein-rich diet not only helps with dopamine production, but also offers the side benefit of suppressing high blood sugar.

Take mineral supplements, specifically magnesium. Most of our foods unfortunately lack the nutrients needed for a healthy-functioning body, and we may need supplements to reach optimum levels of mineral intake.

I've left my favorite for last. Dopamine is released when we feel we've accomplished something and deserve a reward. The good news is, that you don't have to earn an Olympic medal, or colonize Mars to qualify — even simply washing your car will do — as long as that was on your to-do list and it gives you the pleasure of having one less thing to do. To-do lists have more benefits than just organization of the day. In other words, keeping a to-do list can give the benefit of satisfaction of getting things done. As you go through these solidified goals, crossing each item off as completed can give you a sense of accomplishment. The resulting pleasure is likely caused by dopamine release. Turn your daily tasks into a to-do list, and as you attack and cross them off, one after the other, you'll trigger the release of dopamine to the reward center of your brain as a result of your own satisfaction and acknowledgment of your efforts.

Similarly, having a hobby like playing music, gardening, or woodwork (really, anything that leaves you proud of a tangible result or that imparts a feeling of accomplishment), triggers your brain to release a splash of dopamine to say: "well done, now let's do it again."

4.6.3 — Endorphins

Before I move on to endorphins, I'd like to make a distinction here between hormones and neurotransmitters. Hormones flow through the circulatory system in our blood, while neurotransmitters flow through our central nervous system.

Endorphins, serotonin, and dopamine are neurotransmitters known as the body's natural opioids, as they're received through the same receptors as opioid drugs. The first time I ever heard of endorphins, they were attributed with bringing on the phenomenon known as "runners' high." What they do is provide a pain-blocking effect during strenuous exercise. There are more than 20 endorphins, and low levels of these in the brain are associated with a variety of health problems, from alcoholism and intensified effects of aging, to diabetes and Alzheimer's Disease.

The limbic system, as discussed in earlier chapters, is the center of our emotions and memory. It turns out that it's one of the areas of the brain that's full of opioid receptors, and so is very sensitive to endorphins. In that regard, endorphins are not only pain blockers, but they're also released after pleasurable experiences like having a satisfying meal. Their release is initiated by the hypothalamus and they're produced throughout the body.

One of the differences between endorphins and the other neurotransmitters, serotonin and dopamine, is that the physical sensations associated with endorphin release are much more similar to euphoric states, as in 'runner's high.' Unless one's extraordinarily in touch with their senses, it's unlikely they'd be able to tell when serotonin is flowing through their nervous system, but in my personal experience, endorphins can be detected.

Because of the relationship they have with pain management, deficiency of endorphins can cause chronic pain, migraines, headaches, or even fibromyalgia. A person who's ceased taking pleasure in life anymore, or they're over-emotional and tearing-up easily, or if that person is just having a hard time sleeping — these are signs that the person may be exhibiting a deficiency of endorphins. Have you ever felt sad and emotional, didn't sleep well the night before, and then paid a visit to the mall for a distraction the very next day just to boost your mood? If so, you may have been shopping for endorphins, and made an impulse buy to get a quick fix.

Low blood sugar (which is the description of hypoglycemia), can cause low levels of endorphins. That's why when you find yourselves craving for carbohydrates, sweets, or chocolate, the culprit could be low endorphin levels.

A physical or emotional trauma can also lead to a vicious cycle of insufficient endorphins. To process traumatic events effectively, we need endorphins, but their secretion can be inhibited by our processing of the very event we're trying to heal from. If you feel lonely, depressed, unmotivated, fatigued, or are experiencing a lot of physical or emotional pain, these are interrelated emotions. If you're able to improve one of these, others will likely resolve as well, because of the chemical interdependence of feelings. But the opposite can also happen. If you're deep in a hole, that hole can easily perpetuate the feeling of being in a deeper hole every day.

So, what can we do to improve our levels of endorphins? Exercise helps with the production of other neurotransmitters as well, but specifically with endorphins, exercise is the magic bullet. The more strenuous the exercise, the more pain there is to resolve, and the more endorphins are produced and released. In my personal experience, nothing shakes off the blues like *intense exercise*. If you don't want to get your heart rate up, then do stretching exercises or yoga. At this point, you're probably able to guess I'm about to suggest meditation for this deficiency as well. And you'll be right. It's striking to see, how the same healthy habits have the power to improve so many aspects of our physical and emotional wellbeing.

Eat healthy + Exercise + Socialize + Meditate
=
Compounding positive effects
everywhere we look

It's an interesting fact that chili pepper can also trigger endorphins, and perhaps that's one reason as to why the cuisines of so many nations feature very spicy dishes.

We've all seen people who have covered their bodies almost entirely with tattoos. Endorphin release is certainly working its magic in this form of body art, as the body responds to this self-inflicted type of pain. Even for a member of Yakuza, sooner or later there's only so much of this type of art one can have done to their skin. But a similar effect can be had with less commitment of time and money with a deep-tissue massage, which can also trigger serotonin and dopamine.

I used to say rather begrudgingly, that anything that's good for you has to taste or feel bad. This, of course, is incorrect. The teenager in me who thought this way didn't realize that eating dark chocolate has its health benefits while tasting fantastic at the same time. It's full of antioxidants, and is a great way to boost neurotransmitters. The same is true for listening to music (such as any type that improves your mood), or dancing, which is probably the single most effective way to encourage these chemicals to flow in the body.

One criticism I have for the American public education system is the lack of emphasis on classical music. When you listen to an orchestra playing a piece from Mozart, Beethoven, Vivaldi or Bach, when you hear the different melodies coming together forming a masterpiece, your brain, as shown in MRI scans, literally glows with joy. Please bring good music into your lives if you haven't already, and dedicate the time to listen to it *intensely*. I'm not referring to something you do in your car, or while folding laundry, but rather while your eyes are closed, and you're deeply immersed in it.

The key here is to avoid sad songs. I don't suggest eradicating them from your life altogether, but if you want your endorphins to flow through your nervous system, sad songs may have the opposite effect. As with the other mood-enhancing things we've touched on in this chapter, the strength of music's ability to influence mood can work in either direction.

4.6.4 — Oxytocin

If you've been looking for a love hormone, this might be the one for you. Oxytocin is produced in the hypothalamus and is stored in the posterior pituitary gland. Release of this hormone enhances the intensity of feelings associated with love, affection, and satisfaction.

For centuries, thinkers and philosophers argued over whether romantic love, a mother's love for her child, brotherly love, and divine love were different feelings. Some argued that they were not related, but many sages across different cultures discredited that conclusion. If the scientific knowledge we have is accurate, sages were right.

Oxytocin is released during orgasm, also during lactation — bonding mothers to their newborns, and even during the everyday act of giving and receiving a hug. In men, testosterone production is improved in the presence of oxytocin. The release of oxytocin also has the effect of enhancing trust in social interactions, and increases feelings of generosity between people. The more oxytocin one has in their bloodstream, the more is encouraged to be produced. As a result, an existing trusting, satisfactory relationship can pave the way to forming new similarly satisfying relationships. Our brains operate within the effects of many feed-back loops, and the web of thoughts, feelings, and emotions, are all very interrelated. Nothing is ever experienced as isolated in the human mind, and perhaps this is what has led Zen teachers to say: how you do one thing, is how you do everything.

Since we've called it the love hormone, it might be obvious to mention, but oxytocin also contributes to the satisfaction to be realized in a romantic relationship.

Do you have a pet? Do you enjoy hugging them, and does that make you feel happy? This might be because petting an animal, especially close physical contact such as giving that pet a heartfelt hug is a clear oxytocin trigger.

So, what does oxytocin have to do with overspending?

We all need to feel connected, validated, and bonded to a group, or a person because we are social animals. The feeling of being part of, or belonging to something bigger than ourselves, is largely associated with oxytocin. From a purely simplistic point of view, it makes sense for a man to buy an expensive car to impress his future mate, as this in theory, would give him a competitive advantage over others driving cheaper cars. This would be a sign of financial success and security. But also, from another purely simplistic point of view, bluntly speaking, this could be financial suicide.

The same argument can be made for giving unaffordable gifts to one another. A large portion of consumer spending in the US happens during the Holiday Season, from October through December. Part of this activity has to do with chasing deals, but the majority of it is shopping for gifts, which many people can't afford. This can be measured in the annual jump in credit card balances during this season.

Oxytocin gives us the feeling of being connected to others. It helps build social bonds and friendships, and helps us feel part of something bigger than ourselves.

Buying a big house and fancy cars to impress our neighbors and relatives, going on vacations beyond our budget to be with a certain set of friends, getting an expensive ring to secure the approval of the desired mate, are in reality, just us chasing oxytocin and other hormones. All of these contribute to a huge price to be paid in order to secure our place in a social group.

Consider for example, the illusion we're sold in the case of the purchase of a diamond. We pay thousands of dollars, for a tiny piece of rock, that is *artificially decreed* by others to be valuable. If all the world's diamonds were to hit the market on

(Godspeed by Edmund Blair Leighton, 1900)

the same day, the price of each would plummet to a fraction of its theoretical cost. But the supply of diamonds is controlled to keep their price high, so that the next gentleman kneeling to propose, will also pay thousands. I suspect that if a group of aliens were to watch this ritual, they'd be truly amazed by how powerful social pressure can be in our species.

In short, there's a cost to be paid for the feelings of being connected, accepted, and validated. Now that we're aware of these, we can look for ways to get those same benefits either free, or at a significant discount to avoid such a financial loss.

So, similar to the other hormones and neurotransmitters, the idea here is to understand the ways to produce oxytocin naturally, so we're not so easily lured into the trap of shopping and overspending in order to receive the same effect.

I'll start with my least favorite method, because so many are overdoing the time they spend on it, but if used correctly, social media can improve oxytocin levels. Seeing pictures of your friends or relatives, sending a note on their birthdays, getting into a fun discussion about your favorite sports team, are examples of how you can feel connected to people even when they're thousands of miles away.

The important principle to remember, is that the relationship between biochemicals and thoughts, feelings, and emotions, has a compounded effect. The more oxytocin you have in your system, the easier it becomes to form new connections. The more connections you form, the more oxytocin you'll be able to produce, and this may create a positive feedback loop. The key is to refrain from overdoing it, potentially leading one to prefer social media to actually spending quality time in the presence of friends and family. Social media, like most things, can be a valuable tool, or another detrimental addiction.

It's a familiar movie scene in which a lover dims the lights, puts smooth music on, lights a candle, and sets the stage for an intimate moment. A relaxed environment, with lights easy on the eyes, and smooth music in the background can bring your mood up, and help you bond with another. We're the product of our environment, and a cozy environment does help create a comfortable feeling. Being comfortable means feeling safe and with fewer stress hormones in our system, we open our receptors to dopamine, serotonin, endorphins and oxytocin.

You don't expect to pick up a pen and start writing with your opposite hand, because you know that you haven't built the necessary muscle memory there. Similarly, some of the solutions being mentioned here may not work the first time, but practice will make perfect. The trouble is, if at first some of these solutions fail, a retail alternative could be chosen as an attempt to compensate. For example, a father could give an expensive gift of toys to make up for his absence on long business trips. Spending quality time with a loved one may require some effort, but that comes at no cost. Trying to compensate for a lack of quality time with gifts requires no effort, but comes with a price tag attached.

Do you want to know why there's such a strong bond between members of an army unit, or a sports team? It's partly because they've shared experiences that stirred strong emotions, which may have felt risky, or adventurous at the time. Sharing these moments with other people is known to increase oxytocin production, which contributes to the forming of strong bonds. This is the result of our brains' defense mechanism to reduce stress and stay focused on the task at hand.

Joining the army may not be an option, but you can join a sports team, or go on an adventurous trip with people you'd like to spend time with. This is one reason you won't forget that "special" trip with your friends. The heightened feelings and emotions during that trip ensure that you were releasing plenty of hormones and neurotransmitters, and your brain's memory recorder flipped its red recording light to "on."

A known oxyctocin-producing activity is giving hugs. Have you heard of a six-second hug? It turns out that after giving a hug for six-seconds or longer, the hypothalamus starts signaling the pituitary gland to produce oxytocin. Try it yourself. Give a long hug to your pet, spouse, child, parent, or friend, and see if this is also true for you. It may take more than six seconds to start feeling it. But each time, that period is going to get shorter and shorter as your body starts recognizing the pattern, and you'll get better and better at it as a result.

Basketball players give each other a high five quite often in a game, including at the free-throw line, because it increases oxytocin and decreases stress, for both parties involved.

Do you want to feel good? Make someone else feel good by giving them a fist bump, a firm handshake, a six-second hug, or affirmative statements. In fact, if this becomes a ritual, happiness hormones begin to flow even during the anticipation period we discussed before.

For the most part, our lives are comprised of a collection of consequences that emerge from our choices, which were the result of our decision-making processes. We need to understand these processes, algorithms, mechanics, and the biochemistry behind it all. Once that's accomplished, we can organize our thoughts and emotions, and reshape our unconscious axioms in such a way that will lead us to a happier and better life, which includes financial stability.

4.7 — Technological Pathway to the Lizard Brain

Now that we've covered the basics of how the human brain works and how it affects our decision making, let's look at a particular industrial sector which uses this information so well against us by manipulating our decisions: technology.

If you're able to control one's thoughts, you can control their actions, preferences, and even their sense of self. This is surprisinglyeasy,especiallywhenwewillinglygivethenecessary information into the hands of tech wizards to do just that. If you give someone the power to turn your attention to any area they choose whenever they choose, relatively quickly, they'll also be able to shape your thoughts, opinions, and eventual actions.

Let's say you want to learn about the Israeli-Palestinian conflict. What would be your first inclination — to go to your local library? I doubt it. You'd surf the web. Then depending on the links presented to you on the first page, after reading a few articles, you might consider yourself sufficiently knowledgeable on the issue. Also, you'd likely have adopted a pro-Israeli or pro-Palestinian position in the process. *Now, who decided which links would populate that first page you were served? You should be aware that they might be determining your opinions on this matter as a result.* Taking this a step further, most likely the political party you'll vote for in the next election will have been somewhat colored by this, as it's a hot topic affecting people's choices at the ballot.

This is what happened during the Brexit referendum in the UK, and the most recent presidential election here in the US. Facebook sold Cambridge Analytica personal traits data on millions of users in both countries, and based on this information, the users were fed specific content in order to steer the political choices of those users to favor Brexit and to favor a Trump presidency. Judging by the results, it is fair to say these tactics were quite successful.

The Cambridge Analytica case is even more troublesome, because the Facebook users who received the manipulated content, tailored to ensure a political campaigns' victory over another, hadn't even inquired or initiated a search, but were served it anyway. Users had no idea what they were receiving was targeted, manipulated content and that this was all happening behind the scenes.

An unlimited number of examples can be given, but in the age of social media feeding us algorithm-driven content, to "help" us decide which posts or news items to follow, I don't think there's much doubt or a reasonable counter argument to be made here, other than to say that: the players are for-profit organizations. What did you expect? YouTube decides which video content to serve up in response to the searches of 2 billion users, and 70% of them will click on curated links. Billions of people are watching what YouTube is serving. This is where we are today, and it's naive to think that in the face of all that, we're free thinkers and not that easily persuaded.

Tech companies don't do all of this because of a worldwide conspiracy, or because they're evil. Their incentive is profit-making and they're responsible to their investors. Your attention is quite profitable to them, and *that's the name of the game right now — who will get most of your attention — because that's what continues to generate advertising revenue.*

If you don't think you're susceptible to manipulation or persuasion, I suggest you think again. In the following chapter on behavioral finance, I'll introduce mental biases and fallacies in more detail. There you'll see, that it doesn't matter how smart or educated you are, we're all susceptible to these errors because of the way the human brain has evolved. Dan Ariely[26,27] in his book *Predictably Irrational* points out that even if you're told beforehand that it's pretentious, being complimented on your jacket is still pleasing. We all have a soft spot for flattery and social approval. In my experience, validation is the number one reward we humans seek from others, and social media platforms take full advantage of this.

Yuval Harari says that we are at a point where "they" know us better than we know ourselves. Speaking at Davos 2020, the World Economic Forum, he gave the striking example of how he didn't know that he was a homosexual until age 21, and a simple screen camera that tracks eye movements could have helped him figure this out in a matter of minutes as he hit puberty. Now imagine, by tracing one's clicks online, especially social media, Facebook knows about a person's preferences, before it's even apparent to them. That's how asymmetric our relationship with technology has exponentially become.

All this know-how is developed to retain our attention, and advertisers continue to improve the effectiveness of their marketing campaigns. How do they do it? Here's how the content shared previously in this chapter comes in handy: we know now that the marketers are targeting and speaking to our lizard brain and amygdala. Unfortunately, due to the current state of evolution of our brains, we react a lot more quickly when an impulse triggers outrage, sex, or fear responses, so every company vying for our attention is going to exploit that path to optimization. This is one reason why conspiracy theories are a lot more widespread today. A certain percentage of people allow themselves to be drawn down a rabbit hole following a dark, scary, or outrageous headline on a social media post, rather than take the time to read an article that is reporting facts. One is addicting, the other requires effort.

Consider these three hurdles we must overcome when navigating in this current digital landscape, where carefully-targeted marketing messages are served to us everywhere we turn, as we search for information. The end result is that it's harder than ever to overcome the urge to buy products that we don't actually need:

1) We have an enormous appetite for social approval.

2) The huge volume of data that advertisers collect on us by tracing our online activities, and the way this data is relentlessly used on us, means that they can hone in on ever-better ways to manipulate our behavior.

3) The biochemicals produced in this process, such as dopamine, can be addictive.

This is even more of a threat for the younger generations, specifically millennials and next. Being a Gen Xer myself, I still enjoy picking up a paper magazine, or a book, and I do find myself tiring of screen time. I could turn on a dime and follow news on print media if I wanted to, because my mental processes are compatible with this. But millennials and younger generations don't know an alternative to the electronic screen. Their minds have been shaped by this technology, so they are even more susceptible to sophisticated persuasion techniques to target, hook, and exploit them.

So, are we defenseless to these attacks upon our lizard brain? No, absolutely not. If there's one point I'm making in this book, it is: *do everything you can to elevate your mental processes to the higher level of your neocortex, increase your capacity to reason and follow simple logic, and all aspects of your life will improve — including your personal finances.* There's a reason I devote a short chapter to mindfulness and meditation, and that reason is: *these two practices work like magic.*

If you become aware of how your lizard brain and the limbic system works, and how it is being hacked, then you can start looking for ways to install the necessary firewalls. If there's one teaching that is foundational in almost all disciplines, it is:

Know thyself

If you're seeking a defense mechanism to counter modern life's challenges, knowing what your true wishes are, and where you want to go in life, should be on top of your list. The surest path to that end is a personal practice of meditation.

When Alan Watts was asked if there's free will, he brilliantly answered:

There is free will to the extent that you know who you are.

We all have to demand that our lawmakers do more to protect us from how we're targeted on the Internet. They should increase legal protections of our privacy and of our rights to the ownership of our personal data — especially when it comes to marketing that's targeted to young adults and children. To argue that each person should be responsible for themselves, is equivalent to saying that we each should be able to catch the magician's sleight of hand and act accordingly. This is useless as a protection, especially since our neocortex, which would be responsible for exposing the magician, isn't even fully developed until the age of 25.

Wearing seat-belts is mandatory to save lives, and we have a government agency that's set up specifically to ensure the safety of food and drugs circulated in the US. Why should protections be any more lenient for what our minds are fed?

This is also true for corporations. We do vote with our dollars and they're definitely listening. So if we were to support only companies committed to operating a fair playing field, we then could begin to create our own preferred reality. As of now, our reality, curated by others, is not necessarily serving the average person's best interests.

Next, we'll embark upon the hot topic of the psychology of money and finance. The rational highways we built and strengthened by becoming aware of how the brain works, will help us see the truth behind fallacies that we'll encounter in the chapter coming up. These two bodies of information — Brain function and Behavioral Finance — go hand in hand. We can build mental muscle memory conducive to better decision-making if we're aware of the hardware and software behind our actions. Behavioral finance is an important piece of the puzzle to lead us in that direction.

Principles of Behavioral Finance

5.1 — Introduction

Nobel Laureate, economist, and Professor Robert Shiller[24,25] describes the field of behavioral finance, or more broadly behavioral economics, as: "...a revolution that has occurred in finance and economics over the last 20 or 30 years." It's a relatively new field studying the psychology behind investor and consumer behavior. It attempts to rationally explain the irrational choices that humans tend to make, especially around money and consumption.

Traditional economics starts with the broad assumption that consumers, investors, and that people in general, are rational beings (whose decisions always seek maximum gain with minimum risk). This axiom is embedded in most of the major economic theories and yet, we now know this is not true. In fact, more and more social studies are finding that human behavior is far from being rational. I'm guessing that's why Robert Shiller called behavioral finance a "revolution," as it challenges the entire field of economics' findings. We make choices irrationally, and our rational minds work around the clock to justify these irrational choices. This claim needs to be tested, and the name of this process is behavioral finance.

5.2 — Fallacies and Biases

The average human brain hasn't evolved to process complex numerical variables. Prior to the Industrial Revolution, this weakness didn't cause significant problems, as we weren't presented with quantifiable uncertainties on a daily basis requiring a decision be made quickly.

Once goods and services were able to be produced and distributed on an ever-increasing scale, then two variables started to dominate our decisions around money: gains/losses and the probabilities of each. Daniel Kahneman[24,25] and Amos Tversky called it the "Prospect Theory," and added the terminology of: "value function" and "weighting function." Our incompetence with accurately gauging both the value of our gains and losses, and the probability of certain events to occur, has been widely studied and documented. In simple terms, we're just not good at computing the chances that any particular event will occur, and also, whether its occurrence will present a positive or negative outcome.

The improvement, in most of the mental processing errors discussed within the context of behavioral finance, relies more upon education than a higher IQ. We're not helpless, and the remedy is awareness. So, in the following sections, we'll look at the cognitive and processing errors that most of us are prone to make, and discover ways to improve our decision-making processes. As you read on, consider taking a moment after each section to identify and reflect upon times you've fallen victim to a similar type of bias or fallacy. That way, you'll find the information easier to recall for your future use.

How is it that we have biases in the first place? Because unless we have a trained mind, which should be the ultimate goal here, we tend to prefer feeling good about the results of our mental processes, over seeing the uncomfortable truth. An easy way to feel good is to tell ourselves what we want to hear, including that we were right about some choice or decision. If we can't convince ourselves we were right, then at least we can ignore the fact that we were wrong, which is good enough for the cozy feeling we're after. As a result, more of our energy is concentrated on finding ways to conclude that we were right, versus seeking the most optimum solution to whatever problem or question is before us.

This is a deadly recipe for money/investment management. The most successful investors are those who never invest emotionally, those who have no attachment to their choices, and so are able to switch sides instantly. This is why it can be a challenge to manage your own investments — it's no longer an abstract dollar amount that's potentially at risk, but your own future dreams. In the face of that, it's a lot more difficult to manage your thoughts and emotions, and remain objective.

Our brain doesn't want to work any harder than it already has to, with the staggering amount of internal and external information it must sort through and make sense of. So it oversimplifies messages, creates shortcuts, and in general, looks for an easy way out. I call this the "Lazy Option Bias," and it doesn't bode well for us when this bias is in play in modern life, especially not with personal finance.

I'll start this discussion with my favorite bias. When I started learning about behavioral finance, this particular one caught my attention, because I found that I myself tended to exhibit it all the time. As my wife will cheerfully tell you; I like being proven right, (admittedly not my most attractive trait), and confirmation bias is an easy path to that end. So let's begin.

5.2.1 — Confirmation Bias

If you're a republican on the political spectrum in the US, your preferred news channel is probably Fox News. Conversely, a democrat's choice is likely NPR, MSNBC, or CNN. This is because we like to hear about the information that supports and confirms our existing biases, hence the name given to "Confirmation Bias."

Let's imagine you've decided to get a European car because you like both the way they look, and the prestige you perceive that they bring to their owners. As a conscientious consumer, you'd still like to do some research before you buy one, so you talk to your mechanic about it. Unfortunately though, he warns you about the expensive maintenance routine many European cars are known to require. He suggests that a Japanese car might make better sense for a middle-income family like yours, as they're also known to be more reliable.

Well, you didn't particularly like that response, so you call a European car dealer to learn more. They're excited to inform you that your mechanic didn't really know what he was talking about, because after all, European cars are generally safer, and the driving experience is so superior. Now *that* is what you wanted to hear.

You also called a friend who owned a European car himself, and he told you that he was very happy about his choice, which by the way, could be the result of another bias we'll discuss a little later.

So you're now on a mission to justify your decision to buy a European car, even though it is above your means, and even though an expert has informed you that it's not the best choice for you. Why would you go out of your way to do that? Because your mechanic has caused you to experience a discomfort commonly known as "cognitive dissonance," which is a type of frustration caused by having two different motivations that pull you in different directions at the same time, on the same subject. A part of you wants to do the right thing and listen to your mechanic, an objective expert. Another part of you wants to listen to your friend, and stick with your initial conviction. Confirmation bias helps you resolve this, but not in the right way. It will cause you to subconsciously ignore the information that is against your biases, and will encourage you to focus on that which confirms your beliefs. In this case, your brain will help you choose every article, video, and commercial, which shows you what a good decision buying a European car will be. If left unchecked, you'll feel good about it, and you *will* buy it.

The right way to solve this problem would be to objectively evaluate the facts, and exclude your emotions. For instance, you'd call your mechanic, and with his guidance, you'd determine what the extra cost of maintenance would actually add to the real cost of owning that European car. If you did that, you'd either find it palatable, or you'd find it alarming, and your choice would be clear. Discussing this with your spouse, asking your financial advisor, and trying to see clearly if this purchase will help or hurt your other financial goals — such as education or retirement — is the rational thing to do.

I see confirmation bias in investors all the time. People who believe the stock market will go up, search for clues that will confirm this, so they'll feel good about their decision. On the very same day, people worried about a downturn, will look at it the opposite way. The thing is, anyone can easily, at any given moment, make a case for bullish or bearish decisions:

Bullish during an uptrend:
The market is going up, the trend is intact, you don't want to miss it. The trend is your friend, so we should be buying.

Bearish during an uptrend:
The market has gone up too much, and a correction is due. Asset prices are inflated, so I advise taking some profit.

Bullish during a downtrend:
The market's going down, but it"s only paper loss.
You don't want to crystallize your losses. You make money
when you buy at lows. These are good prices to buy.

Bearish during a downtrend:
It's time to manage risk, cut losses, limit our downside
exposure, take some gains off the table.

You see, in every market, a move in each direction can be
easily justified if you look for information that suits you. So
how can we avoid confirmation bias? Start by becoming
comfortable with these two statements:

1) I don't know.

2) I was wrong.

and also, these two questions:

1) What am I basing my decision on?

2) Have I considered all the facts objectively, especially
   those of the opposing side?

For instance, if you think the market will go up, do spend some
time looking for bearish evidence and vice versa. Consciously
seek out and evaluate opposing views that cause you to have
that uncomfortable feeling of an internal conflict, where the
balanced and better-informed decisions may well emerge.

Catch yourself when you feel the need to be proven right,
because it could be a sign of your own doubts about your
convictions. If someone told you that you had a bug on your
tail, you wouldn't even bother to address this, because you'd
be more than a 100% sure that you don't have a tail, let
alone a tail with a bug on it. You'd see no reason to prove it.
Ask constantly, what is more important to you? To be right, or
to make the best decision based on the information available

at the time? If you lean toward the former, tough years may be ahead. But if you seek the latter, being aware of this bias will be very useful.

In short, keep skepticism near and be aware of the tendency to overvalue information that confirms your existing bias. Look for opposite views to balance each other out, and in general, leave no stone unturned in the face of making an important decision.

Hegel was a huge advocate of learning from the opposite view. He wrote that the best ideas would come from the synthesis of opposing views on both sides, and thus it's important to give both sides equal weight. Since we know our own existing biases well enough already, we'll now spend some time getting to know some counter arguments, and trace the path of their synthesis.

5.2.2 — Dunning-Kruger Effect

Psychologists David Dunning and Justin Kruger, in a study published in 1999, identified that *people who know little, or next to nothing about a topic, feel more confident* about their abilities, and have an overinflated sense of competence in that subject, compared to those with great knowledge and expertise in it. The unskilled person overestimates their grasp of a subject because of their ignorance regarding the magnitude of what there is to know about it. While conversely, the expertly-skilled person underestimates their own expertise, because of a tendency to overestimate what others know about the subject. This is the general definition of the bias known as the Dunning-Krueger effect.

When we begin to learn about, for instance, how to fix a bike, we don't even know what we don't know, so it may seem like an easy task to accomplish. At the end of the day, how hard can it be? After all, it's just a bike. But as we start learning more and more about it, we'll realize how many little details there are, and even more importantly, how much time and commitment is needed to practice the necessary skills to gain competence with it. If you know nothing about fixing a bike, you may decide to give it a shot. In the end, why not? But if you took a one-hour course on it, you'd then be informed enough to realize you should either enlist an expert, or learn more about the subject. Because if you don't, your brakes could fail, and a serious injury might be in your future.

Years ago, a psychologist friend and I were chatting about an incident between two people. I quickly concluded that one of them was "crazy" (not a proud moment, but that's what happened). I was also pretty confident in my assessment. My friend, who has a doctorate in clinical psychology, wasn't so sure about it. In fact, she said to me "Buğra, I've spent years studying human psychology, and I know I haven't even begun to scratch the surface of it. As a result, I know enough to say that I've got no idea what's going through that guy's mind. You, on the other hand, are clueless and yet so sure about your conviction." She was rightfully frustrated with me, because that was exactly what was happening. I was being arrogantly ignorant, and worse, I was being stubborn about it — something I now watch for in myself, and something that I now see in others all the time.

And what's the cause of this? Yes, it's the Dunning-Krueger effect. Going back to the bike repair example, if a beginner at repair were to read through a checklist before declaring a bike safe to ride, completion of this task alone would uncover this cognitive error. On the flip side, if a true expert in the field of bike repair were to become privy to what others had scored in a test of the same subject matter, they'd feel much better and more confident about their own level of competence. The key is realizing that this effect is a cognitive error, not an ego trip, or a psychological problem. It can easily be addressed by having more awareness of the effect, as in the examples given. In many cases, it's simply a common bias that we'll do better to be aware of.

How is the Dunning-Kruger effect harmful to your finances? Most people think they don't need to consult with an expert, and that they can file their own taxes, manage their own investments, write their own will, and do their own accounting. This may very well be true for some people, especially if they have experience in a related field. But personal finance is a complex field. There are so many products, strategies, frequently-changing tax laws, global and domestic economic developments, and political developments as well, that it's a full-time job that requires continuous education.

I also see situations where clients trying to avoid paying consultation fees to an insurance specialist, may end up over-insuring for some risks and under-insuring for others. Or sometimes it's an inattention to tax law that could land these folks with a bill for taxes and penalties for their non-compliance. Studies show that investors in mutual funds rarely receive their fund's portfolio return, because they simply bought and sold at suboptimal times. Men fall prey to this cognitive bias easily. I see significantly fewer female clients who think they could or should manage their own investments, but many male clients who do think they could. The result is usually that men will show an over-concentration of stock positions in their portfolios, and will show a history of buying when prices are high, and selling when they're low. Studies show that women make better investors for another reason: they tend to be more risk-averse, and their focus is on longer-term goals. Men, however, tend to take more risk, and seek shorter-term results).

So then, what to do? Of course, educate yourself in the field of personal finance, but do also consult with an expert when you're ready to execute. In short, don't overestimate your skills or knowledge, and *talk to an expert before you make important decisions*. Even better, talk to more than one expert and compare their recommendations.

I'll leave you with a quote from one of my favorite thinkers, Bertrand Russell (1872-1970):

The whole problem with the world is that fools and fanatics are so certain of themselves, yet wiser people so full of doubts.

5.2.3 — Sunk Cost Fallacy

This section examines an effect that we can certainly all relate to. We've all experienced being served more food at a nice dinner out than we had room for. As we get to where we're full, we know we'll suffer being overfull if we continue, but we feel obliged to keep eating — Why? Because we've paid for it.

In general, we tend to cling on to things and have a hard time letting go of them because: *they're paid for.*

As a result, you may end up with the discomfort of having too much food in your stomach, and complain about being stuffed and uncomfortable on the way home. Later that night, it may disrupt your sleep, and affect your job performance the next day as well. You're already aware of the potentiality of all of this, and yet, you weren't able to stop yourself. The fact is, *how much you've paid for your meal is irrelevant to how much of it you should consume.*

A rational person wouldn't use the cost of the meal that's already on the bill as a factor for deciding how much of the dinner to consume. But it turns out we're not all that rational. This highlights the fallacy in the "sunk cost fallacy." The cost is already "sunk," and therefore it shouldn't affect any future decisions, because nothing can bring it back.

A more illustrative example of the "sunk cost fallacy" can be seen in people's garages or self-storage spaces. Most Americans own plenty of "stuff" they no longer need or use, and they can't let go of it because again: they've paid for it. But keeping this stuff collecting dust, especially in self-storage, actually continues to cost money. Good questions to ask include: when was the last time this was used, or how often has it been used? You can't remember? Then donate or sell it. Get rid of it so you save the space and cost of holding on to it.

In investments, a comparable parallel is when one holds onto securities that are no longer believed to be profitable. But there are better options out there than clinging to them, as they continue to incur a cost in the present as well as all that was spent to buy and hold onto them in the past.

This phenomenon also affects you with respect to the time and effort you've invested or sunk into a process as well. For instance, let's say you've just spent the last hour negotiating with a car dealer, and you were to allow this fact to make you feel like you should purchase the car — not because of the quality of the car, but because of the time you've spent. A similar example highlights how you feel when your time and money has been invested in a bad movie. Because you've watched half of it, you'll likely feel obliged to finish it, even though you're pretty sure you won't enjoy the second half either. Watching the rest will not bring back the time you've lost. You're better off to pivot and find a great way to spend the next hour, independent of the hour you lost on the bad movie.

There's no documented or commonly-known bias or fallacy called "opportunity cost blindness," but if I may, I'd like to take this opportunity to introduce and describe this bias, here and now. Many clients say: "Yes, but I'd like to keep it," not realizing, that the cost of holding that stock becomes a missed opportunity elsewhere. Hope is not a strategy, and what the stock originally cost you is no reason to continue to hold it. Your money could be working much harder for you if you were to invest it on a different stock, (and if you don't sell it, you can now be seen to be blindly paying the cost of that lost opportunity).

So why do we act this way? It's because of sunk cost fallacy, loss aversion, and the endowment effect. These are interrelated, so they make a great gateway to each other in this chapter.

5.2.4 Loss Aversion and Endowment Effect

If I presented you with a deal in which you flip a coin and if heads you win $140, or if tails, you lose $100, how would you bet? Research shows that many people won't play this game, even though the probability of each event is equal, and the potential gains far exceed the potential losses.

The value of the offer is this:

50% (probability of getting heads) x $140 (the winnings) = $70

50% (probability of getting tails) x - $100 (the loss) = -$50

Total = $70 - $50 = $20

But why do studies show that most people would stay away from this offer? Because of loss aversion. Most people weigh the psychological effects of a loss, and attribute it as having much more weight than a profit. In my experience, which I'll share as anecdotal evidence, is that this relationship is nearly 2:1. In other words, people tend to consider a $100 gain to being equal to a $50 loss in emotional weight.

The implications of how this shows itself in our finances and investments is that we either take too much or too little risk. During the earlier stages of life, an investor can afford to have an aggressive portfolio as there's time to recover from market fluctuations and volatility. But as a result of loss aversion, this train, for many investors, unfortunately leaves the station without being capitalized on. Then during the later years of the investor's life, when risk levels should be lowered, to prevent a shock from wiping out their portfolio, investors often take on too much risk, trying to make up for lost years of opportunity.

The reason I made time in the previous chapter to describe brain physiology and its effects on our psychology, was to help us recognize the emotional factors at play as we try to make the right moves. I'm sure many of you are wondering: why discuss brain biology in a book about personal finance?

A rational, neocortex-driven mind would take the coin toss offer presented above any day of the week, because if played enough times, eventually, the odds stack up in favor of the player. But as we've seen, due to loss aversion, this is very hard to do for many people. The only way to avoid loss aversion is to quantify your potential gains and losses and evaluate the probabilities. This may sound like a mouthful as well as something difficult to do. But my advice stands, that when it comes to important decisions about your finances, consider consulting with an expert.

Next, let's discuss the Endowment Effect. We tend to consider things to be more valuable once we own them, and as a result, we have a hard time letting go of them. To us, our cars, furniture, homes, and current investments are worth much more to us than their value would be to others. This is one reason why many people hold on to losing investments instead of replacing them with better alternatives. I can't tell you how many times I've wanted to buy a secondhand item but decided against it, as the owner had put the price up too close to what it would have cost to buy the thing brand new.

Throughout my career, whenever I've gained a new client, there's always one commonality that stands out — and that's the number of losing stocks in their portfolios. It's almost a specialized talent many possessed, to find, buy and hold on to losing investments. Now, of course, this isn't the case, but rather, it was the endowment effect which kept them from cutting those under-performing stocks from their portfolios. A typical response to the question "Why do you hold on to these?" is: "Well, it's just a paper loss, I'm waiting for them to go back up. If I sold now, I'd realize losses, so I'd much rather wait for them to go back up."

I'd be OK with that answer if there were any reasonable explanations as to why and how these would ever recover their values. Usually, though, hope as a strategy doesn't pan out.

So, for your investments, have specific goals, and if they're not met, be the first person to change direction. One of my favorite stories involves a popular economist, John Maynard Keynes (1883-1946). In an argument, when he was criticized for conflicting with his earlier statements, he responded by these famous words, which I always try to remember:

When facts change, I change my mind.
What do you do, Sir?

One of the best intentions in the writing of this book, is that I might convince you, the reader, that you're not your thoughts, emotions, or convictions. You can and should drop them the moment they become out of sync with current circumstances. Your ego will resist, so it's easier said than done. And that's why it's a complex issue that requires a lot of mental training. I don't suggest dropping your values at every turn, I only mean that you shouldn't stick with a wrong decision due to the hope that it will right itself at some point. Your job isn't to always be right, but rather, to just attempt to make the best decision you can with the information that's available to you at any given time. And that includes correcting errors along the way.

5.2.5 — Anchoring

If you'd like to meet master salesmen, who've learned every trick in the book to make you feel bad for leaving their store without buying their product, you should visit a carpet shop in Istanbul, Turkey.

When you walk in, the first thing they'll do is to order you a cup of tea, so you'll feel obliged to sit and invest some time. This sets up the sunk cost bias. While chatting about your trip, you'll begin to more feel relaxed. You'll be happy to be sitting in a charming, air-conditioned environment, with friendly people who speak your language, (let the oxytocin production begin). It just so happens that one of them, has visited or has lived in the US, and has a cousin still living in New Jersey or Florida (this cousin will come handy later). In 10-15 minutes, you'll feel like you've made some local friends, who are happy to help you find the best restaurants, hotels, and tour guides. The mood is set to move to stage 2.

Eventually, you'll decide it would be rude not to look at a rug or two before you leave, which they masterfully display by flipping them in the air, (if the display rug is small enough to flip). What could you possibly have to lose?

The first rug shown to you will be worth $50,000. What a beautiful thing, but who could afford that? Absolutely not! They'll happily share with you that customers from the US have purchased rugs for even $100,000 (which is true by the way). But that wasn't going to be you, not that day.

So, how about a $15,000 rug? That's certainly a much more affordable rug by comparison, and still it looks very authentic and beautiful. But you didn't stop by to drop $15,000 in a rug store! All you wanted to do was to check them out. So, still the answer is, no. All this time, the anchoring effect has been accomplished, and it's time to move to stage 3.

They'll do you one last favor, for you only, a special deal because they really like you, and you know, after all, now you're friends. They'll offer you a $5,000 rug, that they won't even make a profit on, supposedly. It's been sitting in their store because a millionaire returned it, as his wife hadn't agreed with his choice. She had replaced it with a $25,000 rug. They'd be willing to let you have it so you have something to take back home with you. They just don't want you to leave empty-handed, that's all.

If a millionaire chose it at one time, it must be a fine rug. And compared to the others the price isn't that bad. You can swipe a credit card and take it home. But wait! How will you get it home? Shipping will cost another $300-$400 at least, and that's too much! Don't worry, that'll be taken care of during the closing ceremony of stage 4.

Goodness, he'd hate to see you leave without being able to take advantage of such a great deal, and since you've become friends, he'll ship it to you for free. In fact, it turns out that one of them is flying to New Jersey next week to see his cousin (I warned you about that cousin). So they'll add it to their luggage, and ship within the US domestically to reduce the cost.

You can't believe your ears, they'll do that for you? They'll ship it for free? Done deal, show me where to sign.

All the while, after this 45-minute charade, they sold you exactly the rug they chose for you from the moment you stepped inside. And now that they have your address, I promise you, you'll be visited within a year, by someone giving you an opportunity to buy more rugs (bigger and better deals this time too), in the comfort of your own home. It'll be your living room where the rugs will be flying in the air.

Ladies and gentlemen, the above example is a thousand-year-old textbook example of "anchoring." The first piece of information presented, in this case the $50,000 rug, is what influences the perceived valuation of the items to follow.

In a study conducted by Kahneman and Tversky,[28] people were asked to spin a wheel with numbers from zero to one hundred. Then, the same group was to estimate the ratio of African Americans to the whole of the US population. The result proved that the anchoring effect is not only real, but it also can be random and unrelated. People who got 10 on the wheel guessed 25%, and people who got 65, estimated the ratio to be 45%. So with completely unrelated data points, the first number to come up had affected the answers to follow.

How does this affect you? Even if you've no plans to visit a Turkish rug store any time soon (don't get me wrong, you can find great rugs there), you're still being targeted by the same strategy. The next time you get a magazine subscription offer, look at your options more carefully, and under the light of this information. You'll probably have three choices that look something like this:

a) $200 for 12 months
b) $50 for 6 months
c) $9 monthly

In the first option, you're being anchored at almost $17 per month, so $50 for 6 months begins to look pretty good. If you don't like it, you can cancel in 6 months anyway, so this way you'll get a fair chance to test it out. Choices A and C are there only to steer you toward choice B. You think you've considered all the options and have come to a wise decision, when all along, B was the choice the magazine publishers wanted you to choose in the first place.

So, the next time you consider purchasing a product, take a breath, clear your head, and beware of the anchoring effect. As seen above, even random information can cloud your judgment. Also, when comparing your choices, try to limit your comparison to similar qualities and quantities from different suppliers, as opposed to multiple offerings from a single seller.

I have a friend who used to work in a retail store. We'll finish this section with what he shared with me.

The owners of this shop knew that most people were looking for deals during the Holiday Season. And more than 70% of their shop's revenue is usually made during the last 3 months of the year. So, they'd jack up the prices by doubling them from January through September, and cut them in half during the Holiday Season. Because the price tag said 50% off, customers felt like they were getting a great deal, while in fact, that was the shop owner's desired price all along.

5.2.6 — Placebo and Barnum Effect

According to Harvard Health Publishing[29] (Harvard Medical School): "Now science has found that under the right circumstances, a placebo can be just as effective as traditional treatments." If you believe that a cure will work, your brain can convince you of its effectiveness regardless of the contents of the treatment. "They have been shown to be most effective for conditions like pain management, stress-related insomnia, and cancer treatment side effects, like fatigue and nausea."

The effect of placebos is important to consider, because they can play such a powerful role in producing the very neurotransmitters we discussed in chapter 4.

Harvard Health Publishing continues: "How placebos work is still not quite understood, but the process involves a complex neurobiological reaction that includes everything from *increases in feel-good neurotransmitters, like endorphins and dopamine, to greater activity in certain brain regions linked to moods, emotional reactions, and self-awareness. All of it can have therapeutic benefits.*" This hits the nail on the head. And this is why and how religions or other faith-based practices work. If you believe in your horoscope, there's a decent chance you'll turn the reading into a reality. Or if you believe a certain prayer helps you ease pain, it actually can. That may not be because someone is listening on the other side, but because of the placebo effect.

Ted Kaptchuk of Harvard Medical School emphasizes the importance of ritual. "You also need the ritual of treatment. When you look at these studies that compare drugs with placebos, the entire environmental and ritual factor is at work." He continues, "You have to go to a clinic at certain times and be examined by medical professionals in white coats. You receive all kinds of exotic pills and undergo strange procedures. All this can have a profound impact on how the body perceives symptoms because you feel you are getting attention and care."

Speaking of the importance of rituals, remember the story of Baron Godfrey and his son Balian on page 81? Do you think the dramatic ceremony of initiating Balian to knighthood had a profound effect on him? To impart the same effect, could they have simply talked about it, shaken hands and parted ways? The ritual is *invaluable* to the process of preparing Balian for knighthood, strengthening both his conviction and capacity to bear the responsibilities of a knight. The ritual elevates emotions, and as we see from our own lives, we recall memories more vividly when strong emotions are attached.

What does the placebo effect have to do with personal finances? It clearly illustrates how powerful our minds can be in creating our own reality, and oddly, how easily we can be misled. It also highlights how important it is, regarding health treatments, to avoid those which haven't been tested by qualified experts (not only to prevent you from buying untested treatments, but also to ensure that what you do spend your money on, is known to provide an actual health benefit). The placebo effect also provides quantifiable evidence of the connection between your thoughts and feelings, and your physical state. Rather than look to the costly remedies served up by consumerism, it's in your best interest to re-adjust your thoughts and find ways to prefer the immediate physical, emotional, and psychological effects that tested treatments provide in response to common ailments. It's you who'll make the final choice.

The belief that our thoughts can affect our physical reality is what has led to the popularity of affirmations, some of which target improving a person's health. Affirmations crafted to be general in nature, are another way that the placebo effect can be used to our advantage. As long as an affirmation can't be proven wrong, the placebo effect will do it's part to create the perception that positive change is occurring.

There's a reason why rituals are such a big part of all religious ceremonies: they enhance the impact of the message upon the individual. Do you want to have a relaxing day on the beach? Make it a ritual, with blankets, an umbrella, a chair, a cooler with your favorite drinks, and a book. Is it more work? Yes, but so are rituals. Will it be more effective in activating your happiness hormones? Yes. By doing so, will you be less likely to go online to shop? Also, yes.

It's important to understand that what you believe in, with regard to yourself, has a fair chance of becoming a self-fulfilling prophecy. In many ways, you are creating your own reality. This tells us two things:

1) There may not be a clear cause and effect relationship affecting particular outcomes in your circumstances, so you might attribute what's come to pass as having more to do with perceived external causes, when instead, what's happened in each instance could owe some measure of credit (or blame), to the power of your mind affecting these outcomes through the placebo effect.

2) We have to be careful about our convictions because even without a legitimate base, they may end up materializing as a result of the placebo effect. This is especially true if we've formed negative expectations. If our reasons to adopt these ideas haven't passed the scrutiny of science, it may be wise to let go of them, if for no other reason than to avoid this affect. If we have some power to create our reality, we might as well make it as positive as possible.

Let's turn now to the Barnum Effect. According to a 2015 Yougov[30] survey, when people were asked: "Do you believe that astrology can tell you something about what will happen in the future?" It may surprise you that 14% of Americans and 8% of Britons said "yes." In the same study, over double the amount, 30% of Americans and 20% of Britons, said "yes" to this question: "Do you believe star signs can tell you something about yourself or another person?"

These percentages equate to millions of believers, enough to sustain a multibillion-dollar industry. Why do some believe in these claims with no scientific evidence to back them? One explanation can be provided by the Barnum Effect.

According to Britannica,[31] the Barnum effect can be seen in, "...the phenomenon that occurs when individuals believe that personality descriptions apply specifically to them (more so than other people), despite the fact that the description is filled with information that applies to a larger group. People are gullible because they think the information is about them, when in fact the information is generic."

P.T. Barnum (1810-1891) is often attributed as the originator of the famous adage: "There's a sucker born every minute," and that's how the Barnum effect gets its name. Psychics, horoscopes, magicians, and palm readers make use of the Barnum effect when they convince people that the descriptions they're presenting are so unique, that they couldn't possibly apply to others. It turns out there's no direct proof that P.T. Barnum actually uttered this phrase, but there is a long list of possible sources if you search the Internet. For instance, in *The 42nd Parallel (1930)*, author Dos Passos gives the attribution to Mark Twain. It also turns out that across the scores of years there have been many colorful ways to word this descriptive aphorism, though the meaning, having to do with the ubiquitousness of gullibility, remains the same.

But it's likely that the adage was first given Barnum's name by psychologist Paul Meehl, in his 1956 essay, *Wanted—A Good Cookbook*. In this essay, Meehl calls out this kind of gullibility as being a distinct psychological effect. He describes that certain kinds of vague statements about people can be crafted with intentional ambiguity, such that those with the Barnum Effect gullibility, are more likely to interpret the characteristics as particularly relevant to them (though when looked at objectively, they could be applied to anyone). As discussed earlier, confirmation bias looks for clues that will confirm existing beliefs or desired outcomes. For instance, I could say: "During the next month, all those born in January will get a chance to reflect on their past experiences, and will receive a warning from a friend or relative about their business dealings. This month, they should pay closer attention to their professional transactions and use their past experiences more frequently as a guide." And this would be applicable, I'm going to guess, for at least half of the readers. I just made a general enough statement, that many people could relate to. The trick is, it sounded specific enough to be credible. Too vague, and the statement would fail an escape velocity test as it were, and people would catch it under their radar as being too general. Too specific and the statement would be unrelatable.

What's the harm in believing in horoscopes, psychic readings, paranormal activity, and magic?

If it's true that *how you do one thing is how you do everything*, then you might want to start paying attention to how you form your mental muscle memory. Allowing yourself to fill in the blanks so you can believe what you want to believe, could be creating minefields for yourself, especially when it comes to the more important decision-making processes in your life.

Do you want to take charge, and change your life for the better? Then consider taking a closer look at your internal narrative and belief system. According to Immanuel Kant (1742-1804), an influential Age of Enlightenment German philosopher, there's no philosophy to learn, but rather only knowledge of how to apply it. What's meant here, is that philosophy, like science, is a methodology, rather than a conclusive body of information. You can learn how to use the techniques of philosophy (or science) to help you process information in a more productive, effective, and healthy way. As a result, the chances that you'll reach constructive, accurate conclusions more frequently, will increase. If you make it a habit of passing all ideas through the filter of critical thinking, you'll establish the most reliable guide within yourself: reason. The Barnum Effect, trying to see clues about your personal life from information that can be applied to millions of others, is a violation of that journey, and one that can have negative spillover effects. *When there's a bug in the system, it can threaten the system as a whole.* In short, once your internal narrative guides you to see things the way you want to see them, and hear the things you want to hear, then you'll be removed from discerning reality in other aspects of life as well.

Things are what they are, not what you think they are.

The second reason is purely financial. Once what you've learned about the Barnum Effect is internalized, you'll spend less money on psychics, horoscopes, and tarot cards. Some of these professionals charge hundreds of dollars for an hour, and their customers often travel long distances for these readings. These are resources which could be directed to a savings account or used for paying down a credit card balance. At the end of the day, you'll decide what your future will look like, by the actions you take and the decisions you make in the present. Exactly the same way that your past actions have shaped your present reality. There's nothing a psychic can say that will change this. Think critically and clearly, and prefer to see things as they are, rather than seeing them as you want them to be.

5.2.7 — Optimism and Pessimism Bias

If a person believes bad things are less likely to happen to them, they likely have an Optimism Bias. In the case of this bias, a person who drives in an unsafe manner, is less likely to consider the possibility of getting into an accident. If they smoke, the threat of coming down with lung cancer as a result, becomes a low-probability event for them. They're also less likely to consider the risks of bungee jumping because, in their minds, they likely believe that it'll be "OK."

As a result, people with optimism bias take more risks, without analyzing the pros and cons of taking a particular action. So they are, in fact, more prone to accidents, diseases, and other adverse effects stemming from their risky behavior.

This shouldn't be confused with having a positive attitude toward life. Here's an example of the difference: You walk into a bar at 10:00 pm, and the bartender informs you that they'll be closing in an hour, at 11:00 pm. Someone with a positive attitude would see that they've got an hour to enjoy a drink or two, and act accordingly. A person with an optimism bias would imagine that even though 11:00 pm is the official closing time, they could push it to midnight or later because, in their mind, any limitation can surely be bent in their favor. Sometimes, they win this gamble, sometimes not. As we'll see in self-serving bias, an overly optimistic person often quickly forgets the times that outcomes did not go their way. As a result, a person like this may push the bartender too hard for another drink, and end up spending the night in jail. So even though some might think it's good to have an optimism bias, it can often do more harm than good.

In the area of personal finance and investments, this bias results in taking undue risks and exposes a portfolio to possibly irreparable damage, through choosing highly concentrated investments, exceptionally speculative investments, or both. I come across similar cases in client and prospective-client portfolios frequently. When I see that there's a stock or two representing more than half the portfolio, with no exit strategy, I'll discuss the inherent risks of this level of exposure with the client. If there is optimism bias, they'll simply believe that it will be just "fine." Highly concentrated positions can only have two outcomes: huge gains, or losses, (unless an investment that went sideways for a long stretch, which is a low probability). The problem with this equation, is that one huge loss can wipe out all of a client's previous gains, so the optimism bias can backfire just as easily as it might lead to a positive outcome.

Pessimism Bias is the exact opposite of optimism bias. Those with this cognitive tendency will gauge their likelihood to experience a negative outcome as more likely. In their minds, only they get pulled over for speeding, the IRS audits them more often than the average taxpayer, and life is usually unfair.

Unlike the person who doesn't thoroughly check the bungee before jumping, they won't even consider it, as they're convinced that if something could ever go wrong, it's most likely to go wrong when it's their turn.

Unless you're born into independent wealth, to be financially successful, you need to take some amount of calculated risk in your investments. Otherwise, you may be leaving money on the table by not taking advantage of opportunities, and that can lead to falling short of meeting your goals.

Just as with the optimistic investor, I see pessimistic investors frequently as well. It's often an investor who, based on their profile (income, age, and financial goals), should consider taking a higher level of risk to open their portfolio to better potential returns. As it's true that risk and return, more often than not, do go together. But those with the pessimism bias can convince themselves that the market will tank any day now, so they should wait. Unfortunately, for this investor, sometimes years must pass after the bottom of a previous bear market, before they feel comfortable enough to get back in the stock market. And sadly, that could be precisely the wrong time as well, right before the bubble bursts and the next crash unfolds.

Then, what to do? I'll cover this in more detail in the financial management and investing chapter, but to touch on it a bit here, I'd say this: try to stay somewhere in the middle, parallel with your long term objectives and in sync with your investor profile, making changes as needed along the way.

I'll leave this section by reiterating: *hope is not a strategy.* Don't leave important decisions to your biases. Sometimes, when clients ask how I came to a particular decision, my typical answer is: I didn't. Facts did it for me. Put the parameters in place, and let the facts make the decision on your behalf. As facts change, (like Keynes) you must change your mind.

5.2.8 — Actor-Observer Bias, Fundamental Attribution Error, and Self-Serving Bias

In this section, we'll tackle 3 cognitive biases together because they're interrelated, and sometimes feed into the effects of one another.

We'll start with Fundamental Attribution Error. When someone cuts us off in traffic, to what do we attribute this behavior? Most likely, we'll decide that it's due to the flaws in that driver's character, and we'll assume they're a selfish person. It takes extra effort to bring many of us to consider the possibility of extenuating circumstances, such as: "Could it be that they simply didn't see us?" It's a common bias to blame behavior on personality traits. We yell and boo at a baseball player for missing a ball, not knowing that the sun was squarely in his eyes. Why? *Because it is easier to do that.* Simple, but mostly true. It requires a lot less effort to do that, than mustering the empathy to imagine a bigger picture. It's a plain error to judge a situation quickly and attribute the outcome to the related party's personality. This unfair snap judgment is all too often seen when people blindly indulge their biases having to do with a psychological fear or anxiety around "the other," such that based on race, sex, age, sexual orientation, gender identification, or religion, they believe imaginary, disparaging things about that person. In this way, racism and other forms of xenophobia show how damaging the error of fundamental attribution can be.

In reality though, what shapes personal traits? Genetic factors do play a role obviously, but also our environment plays a role, in which case even our personality traits are a product of our circumstances. So we tend put too much emphasis on personal traits, and not enough on circumstances. For the simple reason that it requires more effort to take the processing of this information away from our alligator brain, or the limbic system, and delegate it to our neocortex, we usually punt, and *let our mental muscle memory rule the moment instead.*

The Actor-Observer Effect is similar with one very important twist: *when it is ourselves to be judged, we can spare lots of benefit of the doubt.* When we're the one arriving late, for example, we'll rarely find it due to our poor time management, but instead traffic was to blame, or it was the person who wouldn't let us get off the phone. When *we* miss the ball, we're quick to announce that the sun was in our eyes. Why? Because it's the easier thing to do. If I could coin a phrase here, I'd say that many people suffer from the Lazy Option Bias.

You can imagine that this only applies to instances where we don't want to take personal responsibility, and if instead the result is favorable, we're all too willing to claim ownership of it. This then, is an example of the Self-Serving Bias. *It extends to having a better memory of positive events and being forgetful of negative ones, which is one reason why we have such nostalgia about the past.* This trick of the mind is one of the many ways our egos cope with life's difficulties, and attempt to keep depressive thoughts at bay — by holding on to pleasant memories and suppressing unpleasant ones.

5.2.9 Halo Effect and Reference Bias

The Halo Effect is a cognitive bias that occurs when an individual evaluates an aspect of someone or something as positive or negative, then they *unjustifiably* assume that they must also have other positive or negative traits. The most common application of this is, what we think about a person or a product based on their looks. *In short, this is judging a book by its cover, positively or negatively.*

Studies have shown that we think attractive people are more likely to carry positive traits, such as higher intelligence and kindness. This affects the way we evaluate a job applicant, a political leader, or even an athlete. The way people dress and their general looks can change our perception of them. When people "think" that the person they are speaking to on the phone has good looks, the level of empathy, and benefit of the doubt they will extend to that person increases significantly.

This is also true for products. Consider a bag of potato chips. The cost of the actual chips, is minimal compared to the cost of the packaging. So you're not really paying for the chips, you're paying for the packaging that you'll discard right after eating them. If you think about it, cars of the same size should technically look identical because they should be keeping to aerodynamic efficiency. The reason for so many shape variations has to do with the target market and customer base of the manufacturer, and the known fact that customers value looks over functional factors such as aerodynamics.

Reference Bias, is in many ways similar to the halo effect, in that people tend to judge others' thoughts and opinions by how they view their personalities. In other words, if you like a person, say a politician, you evaluate what they say more favorably. Conversely, we tend to ignore or reject the opinions of a person if we simply don't like them. This can be observed in our daily lives, almost all the time. If you have a cousin you can't stand, you're more likely to dismiss their political ideas, but you'll swing the door of tolerance wide open to your favorite niece's suggestions, no matter how extreme they may sound. *In short, the hallmark of this bias, is that we judge the message, based on our opinion of the messenger,* which is yet another cognitive bias that can mislead us. We may give too much credence to misleading opinions and not enough to valid ones, because of the source.

Again we should ask, "What am I basing my judgment on? Am I reacting to my feelings about the messenger's personality, or can I see the merits or flaws in what's being proposed, regardless of who the messenger is?"

When it comes to investments and personal finance, this bias can be costly. Even for professionals, it's wise to seek and even pay for others' opinions, research, and commentary. If you haven't judged the validity of a proposed idea based on objective criteria, but instead placed the focus on your opinion of the person who developed it, you may end up paying a huge price. Some of these errors should be viewed as tuition cost and can be tolerated. But if repeated, they can cause irreparable damage to your finances.

The appearance or image of the messenger is important to us, simply because humans operate so much based on visual data. Most animals rely instead on their sense of smell, or hearing to save their lives in dangerous situations. Except for raptors and certain insects, we have the best vision in the animal kingdom. From an evolutionary perspective, how our environment, an approaching person or animal looks, or the way an object appears, is always the most important piece of information affecting how we assess our safety. And with regard to how we build our own self-image, we do, what we see. We tend to pick up attributes that we admire in others. Our brain is an excellent machine when it comes to using the copy-and-paste function.

That's why you have to *be* the change you want to see, to demonstrate it to those around you and, hopefully, trigger that copy-and-paste function in them as well. Kids for the most part mirror the behavior of their parents, or others closest to them, such as their friends. This is one reason why parents can get so frustrated when their verbal instructions aren't followed, and you hear the following question: "How many times do I have to tell you?" If they only knew they should have *shown* it instead.

I find myself returning to some important principles repeatedly, and while judging a book by its cover is in line with our minds' operating manual, it's not the best choice. We must overcome lazy option bias, which keeps us from doing research on a product, reading the customer reviews, or computing overall costs, each action of which would require an expenditure of time and energy. We must avoid this mistake: that is making an emotional decision based on the looks of the marketer, and then deluding ourselves by justifying the choice afterward.

Also, from an evolutionary standpoint, our brain tells us that what's familiar, is also safer. A face we recognize is deemed to be more trustworthy in comparison to the face of a stranger. That's why you see a realtor's head-shot photo posted all across town as part of their marketing campaign, so that you'll feel a natural affinity for them when you actually do finally meet them. It'll feel like you already you know them, like them, and you'll also find that it's easier to trust them.

5.3. Chapter Summary

There are so many more cognitive biases that we could go over here, but this is a good stopping point, as I've chosen the most pertinent to the management of finances and investments. If you're interested in investigating these further, by all means do so, as everything you become more aware of, with respect to 'invisible' forces which may color your decision making, can help your financial health.

In summary, there are two types of thought processes:

1) Scientific, fact-based thought processes
2) All the others

The scientific method aims to clear the fog, mute the noise, and helps us see things as they are, not the way we want them to be.

The cognitive biases listed in this chapter are all examples of how our deviating from scientific thought and critical thinking can misguide us at our own expense.

In political science, a popular idea is that every nation gets the political system they deserve. I find it interesting that when people say, "He got what he deserved!" it's usually meant negatively, as a chastisement. Cognitive biases are roadblocks to a better life most of us deserve, so we should be aware of them to help us sidestep these mistakes. To live a better life, we have a responsibility to look for ways to become the kind of person who deserves it — and that's meant to be taken in a positive way. I'll finish this chapter with one of my favorite Goethe quotes:

The hardest thing to see is, what is in front of your eyes.

6
Investing and Financial Planning

As a huge soccer fan, I heard this phrase many times from commentators, coaches, and players: "Soccer's not only about soccer." For fans, it's obviously much more than that. It's part of culture and it figures prominently in our lifestyle. In countries where it's most popular, like Africa, Latin America, and Eurasia, soccer is also business, politics, and a big part of social life. You can tell a lot about a person, if you know what to look for, by their soccer pitch performance. Do they exhibit grace under pressure? Do they take defeat like a champ? Are they selfish with the ball? I guess this is true for many sports.

The same can be said for other daily activities, as in, "cooking is not only about cooking." It's a lifestyle, a way of taking care of yourself and others, saying "I love you" to your friends and family, and a huge component of many cultures.

So, you've guessed where I'm headed here: investing is not only about investing. It's so much more than that. It has the ability to expose your greed on the way up (and fear on the way down), like nothing else can because its results are immediate and will hit you where it hurts the most. You'll quickly find where the boundaries of these two emotions reside in you, and that for the most part they're running the show. Stocks go up because the sentiment is more greed than fear in the given moment, and go down when fear begins to take over.

If you want to learn more about the kind of person you are, investing in capital markets is one of the most expensive ways to do so. A lot of soul searching goes into investment planning, starting with the very basic question, why should I invest? What's my objective? What will be the time frame or duration of this investment? How much risk can I or should I take? All of these are client-driven variables and have a lot to do with the emotional state of the investor.

So far, we've discussed many external and internal factors that can cause us to overspend, and we've looked at ways to overcome the ones most likely to trigger and challenge us without our even being aware of them. If followed, these practical solutions will help you start saving. As you enter your saving phase, this question will arise: "What should be done with this money I've saved?" The information coming in the next chapter will help you make better use of those hard-earned, and even harder-saved dollars.

Spending less than what you earn, (including: paying yourself first, saving, and investing the difference), is the only realistic way you'll be able to reach financial freedom. In the following paragraphs, and with the help of compound interest, I'll demonstrate how even small amounts can add up to a sizable nest egg over time.

I'll introduce basic, as well as must-have considerations, which are vital to building a sound investment plan. Whether you're a do-it-yourself investor, or the client of an investment advisor, you'll want to pay attention to these factors, as it's in your best interest to learn as much as you can about personal finance and investing. The concepts and terms I'll be introducing you to in the coming section are basics, but they're important to being able to understand a wide range of financial environments.

6.1 — Step One: Pay Yourself First

Careful readers will notice this useful trick was mentioned in Chapter 2, and I make no apology, as some concepts are worth repeating. I'll be showing you the math behind this good advice, and how easily it can work for you.

Since we've identified that it's not due to a lack of income, or knowledge, that trips most people up with regard to saving, but instead, it's the lack of executing on a good strategy which causes the disconnect. I'll begin by explaining this simple rule, and it has the power to save you many headaches.

Like many good ideas, it's simple, free of charge, available to everybody, and can be very effective. When you get paid, immediately set aside money in a savings account. You do this — *before* you start spending it — and you choose to make do with what's left. We all know we can get by on less, because for most of us, we've been there at least once. At the end of the day, we must force ourselves to a lower spending threshold, or there'll be nothing to save.

It seems natural to say, we save what's left after expenses. Now let's flip that equation and do a complete 180 degree turn, to get: *We spend what's left after our savings.*

The trick here, is to start small. If you make this out to be a huge mountain, it'll be difficult to climb. Start small, but stick to it. The first lesson is one of self-discipline. If you do one small, correct thing often enough, it will compound to a big thing before you know it. This is also in line with how our brain works, and mental muscle memory will be built. Once you develop the necessary discipline to restrain yourself in this small way, increasing the amount will eventually also become possible, and in time this process will also become fun to do.

Here is a demonstration of this:

You start with paying yourself first, $50 every two weeks, or $100 every month, and invest it in a mutual fund with no transaction cost, no fees to buy in, and with an automatic purchase agreement in a brokerage account. Hypothetically this investment returns on average 0.4% a month, so after a year, your savings will look something like this:

	Starting Balance	Addition	Monthly Return	Ending Balance
Month 1	$0	$100	0.40%	$100.40
Month 2	$100.40	$100	0.40%	$201.20
Month 3	$201.20	$100	0.40%	$302.41
Month 4	$302.41	$100	0.40%	$404.02
Month 5	$404.02	$100	0.40%	$506.03
Month 6	$506.03	$100	0.40%	$608.46
Month 7	$608.46	$100	0.40%	$711.29
Month 8	$711.29	$100	0.40%	$814.54
Month 9	$814.54	$100	0.40%	$918.19
Month 10	$918.19	$100	0.40%	$1,022.27
Month 11	$1,022.27	$100	0.40%	$1,126.76
Month 12	$1,126.76	$100	0.40%	$1,231.66
Total		$1,200	2.64%	

So you may be thinking: that's no big deal, my $1,200 turned into $1,231.66, I made just $31.66 in one year. But keep in mind, your Month 12 contribution only had one month to grow as opposed to 12, that's why your cumulative return is 2.64% instead of 4.8% (0.4 x 12).

But here's what would happen if you were to stick with it. By saving $100 a month for 30 more years, starting with the $1,231.66 which accumulated after your first year:

	Starting Balance	Addition	Annual Return	Ending Balance
Year 1	$1,231.66	$1,200.00	5%	$2,524.91
Year 2	$2,524.91	$1,200.00	5%	$3,882.82
Year 3	$3,882.82	$1,200.00	5%	$5,308.62
Year 4	$5,308.62	$1,200.00	5%	$6,805.71
Year 5	$6,805.71	$1,200.00	5%	$8,377.66
Year 6	$8,377.66	$1,200.00	5%	$10,028.20
Year 7	$10,028.20	$1,200.00	5%	$11,761.28
Year 8	$11,761.28	$1,200.00	5%	$13,581.00
Year 9	$13,581.00	$1,200.00	5%	$15,491.71
Year 10	$15,491.71	$1,200.00	5%	$17,497.96
Year 11	$17,497.96	$1,200.00	5%	$19,604.52
Year 12	$19,604.52	$1,200.00	5%	$21,816.41
Year 13	$21,816.41	$1,200.00	5%	$24,138.89
Year 14	**$24,138.89**	**$1,200.00**	**5%**	**$26,577.50**
Year 15	$26,577.50	$1,200.00	5%	$29,138.04
Year 16	$29,138.04	$1,200.00	5%	$31,826.60
Year 17	$31,826.60	$1,200.00	5%	$34,649.60
Year 18	$34,649.60	$1,200.00	5%	$37,613.74
Year 19	$37,613.74	$1,200.00	5%	$40,726.09
Year 20	$40,726.09	$1,200.00	5%	$43,994.05

Year 21	$43,994.05	$1,200.00	5%	$47,425.42
Year 22	$47,425.42	$1,200.00	5%	$51,028.35
Year 23	$51,028.35	$1,200.00	5%	$54,811.43
Year 24	**$54,811.43**	**$1,200.00**	**5%**	**$58,783.66**
Year 25	$58,783.66	$1,200.00	5%	$62,954.51
Year 26	$62,954.51	$1,200.00	5%	$67,333.90
Year 27	$67,333.90	$1,200.00	5%	$71,932.25
Year 28	**$71,932.25**	**$1,200.00**	**5%**	**$76,760.53**
Year 29	$76,760.53	$1,200.00	5%	$81,830.22
Year 30	$81,830.22	$1,200.00	5%	$87,153.39

What you're looking at here, is the effect of saving and investing on an annual basis as opposed to monthly.

The thing that excites me the most, is what happens in year 14: *your investment return exceeds your savings*, meaning your return is $1,238.61 [$26,577.50 - $1,200 - $24,138.89] and your investment is still $1,200. You have essentially created an imaginary second person saving on your behalf. Now, even though it took 14 years for this one additional imaginary saver to emerge, it only takes 10 years for the next, and 4 more years for the third person to show up to help you out. Meaning in year 28, your return is more than three times your savings rate. How crazy is that? A 5% return will create *3 additional savers for you*, and when you look at it that way, you want to start saving yesterday!

In short, what happened here is that you've saved $37,200 in 31 years and ended up with $87,153.39, which is $49,053.39 more than you put away! You've more than doubled your investment. It's as if someone's given you forty nine thousand dollars for your $100/month savings, wouldn't you like that?

Now, consider this: the current Individual Retirement Account (IRA) contribution maximum amount is $6,000 ($7,000 if you're 50 or older). If you just did that, these numbers would multiply by five, resulting in an ending balance of $435,767, of which $249,767 is growth and $186,000 is principal.

Wouldn't you like someone to give you $250,000 just for being responsible during your working years, with only $500 per month savings? *If you start doing this in your twenties, by your mid-thirties, with a 5% growth rate, your investment returns start to exceed your contribution amount. In your forties double that, and then when you're 50 years old or so, triple it.*

Do you want to get more excited? Realistically, your saving amount wouldn't be stagnant, and would go up by the rate of inflation, if not more. As years go by, typically your earnings potential also improves.

Let's say the inflation rate is 3% and that's by how much you can increase your savings. The progression will now look like this, using that hundred-dollar monthly savings.

	Starting Balance	Addition	Annual Return	Ending Balance
Year 1	$1,231.66	$1,236.00	5%	$2,524.91
Year 2	$2,524.91	$1,273.08	5%	$3,957.84
Year 3	$3,957.84	$1,311.27	5%	$5,501.62
Year 4	$5,501.62	$1,350.61	5%	$7,162.97
Year 5	$7,162.97	$1,391.13	5%	$8,948.98
Year 6	$8,948.98	$1,432.86	5%	$10,867.11
Year 7	$10,867.11	$1,475.85	5%	$12,925.28
Year 8	$12,925.28	$1,520.12	5%	$15,131.80
Year 9	$15,131.80	$1,565.73	5%	$17,495.45

Year 10	$17,495.45	$1,612.70	5%	$20,025.50
Year 11	$20,025.50	$1,661.08	5%	$22,731.71
Year 12	$22,731.71	$1,710.91	5%	$25,624.38
Year 13	$25,624.38	$1,762.24	5%	$28,714.36
Year 14	$28,714.36	$1,815.11	5%	$32,013.10
Year 15	$32,013.10	$1,869.56	5%	$35,532.68
Year 16	$35,532.68	$1,925.65	5%	$39,285.79
Year 17	$39,285.79	$1,983.42	5%	$43,332.67
Year 18	$43,332.67	$2,042.92	5%	$47,547.01
Year 19	$47,547.01	$2,104.21	5%	$52,084.12
Year 20	$52,084.12	$2,167.33	5%	$56,912.88
Year 21	$56,912.88	$2,232.35	5%	$62,049.81
Year 22	$62,049.81	$2,299.32	5%	$67,512.32
Year 23	$67,512.32	$2,368.30	5%	$73,318.77
Year 24	$73,318.77	$2,439.35	5%	$79,488.46
Year 25	$79,488.46	$2,512.53	5%	$86,041.74
Year 26	$86,041.74	$2,587.91	5%	$93,000.06
Year 27	$93,000.06	$2,665.55	5%	$100,385.98
Year 28	$100,385.98	$2,745.51	5%	$108,223.28
Year 29	$108,223.28	$2,827.88	5%	$116,536.97
Year 30	$116,536.97	$2,912.71	5%	$119,526.58

Your ending balance increased to $119,527 from $87,153, a $32,373.19 increase, even though you only increased your savings by $22,803.21.

In short, saving $100 monthly, (considering a 5% annual return, and an inflation rate of 3%), you'd end up with $119,527 in 31 years. If you can save $200/month, you multiply the ending balance by two, and so on and so forth.

As a couple, things get much easier. *Do you want to have a million dollars when you retire?* Save $837 a month for 31 years, that's your goal. That's $418 per person, which is less than the IRA maximum amount of $500 a month (with a 5% return). So, as a couple, even if you stay under the IRA contribution limit, you have the potential to be a millionaire when you retire, if you start early enough. Do you need more reasons to start saving and investing today? If you don't want to be a millionaire, no one can help you become one.

Here's another goal if you want a concrete target to take away from this section: as a couple, start giving yourself a gift on your birthday next year, by starting a plan with a hundred dollars per month savings and invest it in a brokerage account. The following year, increase this amount to $200, and every year after that, increase it by $100. In 5 years, you'll be maxing out your IRA contribution, and will need fewer than 30 years to become a millionaire. For my younger readers, I am hoping that this is a good enough reason to be motivated to stop overspending and start saving.

We've learned about hormones and brain neurotransmitters. Please take my word on this: as you start saving, you'll start feeling good about yourself and your efforts. Firstly, you'll be doing the responsible thing. Secondly, you'll experience the comfort of having a balance in your investment account, which is visible, accessible, growing, and quantifiable. As you watch it grow, it'll give you a sense of accomplishment. Thirdly, you'll begin to feel better about your future. You'll feel safer and more grounded. All of these feelings and emotions are dopamine and serotonin boosters, and once you get the hang of it, you'll be looking forward to paying yourself first. This will evolve from being a headache, an action that pushes you to make sacrifices, to a habit that gives you the opportunity to budget for today, and plan for the future, while feeling good about doing both. By that time, you'll have built the necessary mental muscle memory, and saving and investing will become a part of your healthier, conscious lifestyle. That's the ultimate goal here.

6.2 — Don't Carry Credit Card Debt

A question many financial planners/advisors encounter often is: should I save and invest, or pay off my credit card debt, (or pay down the mortgage)?

I can give a technical answer, and will expand on it as well:

Credit is good...
when the cost of capital is less than the return on capital

Credit is bad...
when the cost of capital is more than the return on capital

Here, the cost of capital is your interest rate, and the return on capital is your investment return. For instance, if you can get 6% on a bond investment, and if your interest rate on a loan is 5%, this credit is good, because you can borrow at 5%, invest at 6%, and keep the 1% margin as profit. This is a simplified illustration to make a point. It gets complicated when taxes, timeline, and risk factors enter the picture, but for now, you get the idea. The return from your investment more than pays the interest on your loan.

This is why and how the rich get richer. They know how to access cheap capital and good investments. When you talk to a businessperson about an investment, the first question they'll have is about the cost of financing. That, along with the projected return, and the risks associated with it, are the three factors needed to make a decision. They rarely use their own funds if they can avoid it.

Now then, why did I start so early in this chapter by saying, "don't carry credit card debt?" Because, for most people the interest on a credit card balance is much higher than a potential return on an investment. In reality, I see only two types of credit cards in the market (taking into consideration that there are always exceptions). One type with more than 15% interest (which you should avoid at all times at all costs), and the other, with a zero percent introductory rate. This second type can be useful at times. They usually come with a transfer fee of 1-3% at the time of this writing, and go up to the usual 16%-22% interest rate range after a year or two.

Since the likelihood of getting a guaranteed investment return above the interest rate of 16%-22% is close to impossible, at least in dollar terms, your payment toward that balance is the best thing you can do with your money. It's the equivalent of getting a guaranteed return on your investment at that rate. So, line up all your credit cards and start paying off the balance with the highest interest rate and work your way down the list.

Alternatively, it can be a good strategy to take advantage of zero percent introductory rate credit cards. Transfer your credit card balances, lowering your interest rate from high double digits to zero (after the initial one-time fee), and do this again before the term of the zero percent interest is over, to another zero percent introductory offer. Here, you're choosing the lesser of two evils, literally. You'd be choosing a 3% one-time transfer fee over the 15% annual interest on your balance. If you keep your payments the same amount, that difference will go toward your principal and will dramatically speed up the paying off of your credit cards.

Another way you can pay off your credit card debt is through debt consolidation. Here, let me introduce you to two products: credit lines and loans. Credit lines work just like a credit card, but usually have a much lower interest rate, especially if you get them from a reputable bank. What you can do is, get a line of credit, pay off all your credit cards, and have only one bank to deal with. This simplification will have three advantages:

1) Problems such as: missing a payment, or having to keep up with multiple lenders will go away. You'll have one lender and one payment to keep track of. In many cases, you can make automatic payments and move on with your life.

2) This simplification will help you keep your focus on more important things. If there's one thing I've learned from successful people, it's that they hate distractions (I know hate is a strong word, but usually that's how strongly they feel about it). If you want stability in any part of your life, you have to focus and pay constant attention to it. Debt consolidation will help you so much with that. It will free up your mind immensely.

3) You'll either improve your cash flow by paying less in interest, or you'll keep the same payment level, (which I recommend), and pay the debt off more quickly.

The next product, which works better if you know exactly how much you need, at exactly what point in time, is a loan rather than a line of credit. The difference is that a line of credit is there for you to use when you need it, but because of that flexibility, the interest rate is usually higher. A loan, all else being equal, will have a lower interest rate, because it will start to accrue interest the moment you sign to start it. It's still preferable though, because of its lower rate. It'll provide you with all the advantages outlined above, with an even more accelerated pay-down schedule. Though again, it'll have less flexibility, because you'll be getting a lump sum to play with, as opposed to a line of credit to use at your discretion. Some lines of credit lose their appeal by charging a minimum annual or monthly fee, which is something to keep an eye out for.

6.3 — Different Kinds of Interest Rates

It's important for the purpose of this book that I describe two types of interest rates. Knowing them can help you become financially free, or conversely, a lack of understanding this can cause you to live in its shackles. This book isn't exclusively on finance, so we'll look at only: simple and compound rates.

According to an urban legend, Einstein called compound interest, "the biggest invention of mankind." It may or may not be fiction, or it may be that Einstein was joking, or being sarcastic. All that aside, I'll invoke the "Never let the facts ruin a good story" rule here, and tell you that he was right. I'm surely no Einstein, but even though I've been serving clients on topics related to personal finance for a long time, I'm still amazed at the power of this concept.

Compound interest is interest on principal and interest. Simple interest is interest on principal only. At first, since interest on interest is a small amount, one might think that it shouldn't make a big difference, but oh boy, it surely does. Let me give you a quick rule of thumb on this topic: it's called the Rule of 72.

If you divide 72 by your compound return rate, you'll find out how long it will take you to double your principal. For instance, if you get 6% compound return on an investment, it will take you 12 years to double your investment, [72 ÷ 6=12].

The detailed progression is below. Your initial investment of $10,000 becomes $20,000 in 12 years. With an 8% compound return, the same result will take 9 years [72 ÷ 9=8], and with 12% it will take 6 years [72 ÷ 12=6], and so on.

	Initial Investment	Rate of Return	End of Period Balance
Year 1	$10,000.00	6%	$10,600.00
Year 2	$10,600.00	6%	$11,236.00
Year 3	$11,236.00	6%	$11,910.16
Year 4	$11,910.16	6%	$12,624.77
Year 5	$12,624.77	6%	$13,382.26
Year 6	$13,382.26	6%	$14,185.19
Year 7	$14,185.19	6%	$15,036.30
Year 8	$15,036.30	6%	$15,938.48
Year 9	$15,938.48	6%	$16,894.79
Year 10	$16,894.79	6%	$17,908.48
Year 11	$17,908.48	6%	$18,982.99
Year 12	$18,982.99	6%	$20,121.96

Now imagine Jack, a 20-year-old, and his cousin Jordan 36. They start saving at the same time, with the same amount — and both with an annual investment return of 9%.

Jack will be able to double his investment every 8 years [72 ÷ 9=8], which will be at age 28, 36, 44, 52, and 60. As a result, his dollars will grow **32 fold.**

Now, please take a moment to wrap your mind around this. *For every one dollar, Jack will end up with $32 because his balance will look like this sequence: 1,2,4,8,16,32.*

Jordan on the other hand, will have $8 for every one dollar at the same age of 60, as his sequence will look like this: 1,2,4,8. He started at 36, 16 years later than Jack, and Jack will be able to double his investment two more times because he started 2 periods (16 years), earlier. Now you know what Einstein was supposedly talking about, and how the rich get richer. It's because they start investing as early as possible.

Let's go back to our initial investment example of the 6% compound return, which showed us it took 12 years to double the principal. Using simple interest rate, it would have taken 16.7 years, and *NOW* you see the difference. In year 24 with a compound return, you'll have doubled your money twice, and your first dollar has become $4. You'd have to wait until year 33.4 to reach the same amount of return using simple interest, and by then, by way of compound interest, you're almost finished with your third doubling period.

This is exactly the same game the credit card companies play, only you're the prey (investment), and they're the beneficiaries (investor). They invest in you, with an 18% return, in which case it will take you only 4 years to double your balance (the amount you owe), using the compound rate [72 ÷ 18=4]! One party's gain is the other's loss, hence the reason why they say that finance is a zero-sum game. The question is, which side do you want to be on? Do you want your credit card balance to double every 4 years (at 18% interest), or would you like to see your investments grow? *It's precisely the same mechanics, and it is up to you, to either fall prey to, or take advantage of.*

To make matters even worse, some credit card companies compound more than once a year. Here is how that looks. Let's say you have a credit card with an 18% or 16% interest rate, which one is better for you? The answer seems easy but it depends on their compounding schedule.

If your card with 16% compounds twice a year (8% twice), a $10,000 balance will grow to more than double that, at $21,589 by the end of year 5. However, an 18% rate on a credit card balance that compounds only once a year will grow to $16,054, or $5,535 less than the card with the lower interest rate. Can you blame Einstein?

I know that this section is a bit more numerical and technical than most people would prefer, so here is a summary to provide you with key takeaways:

1) Don't carry credit card debt unless it's a 0% interest rate and used for debt consolidation.

2) Pay off your credit cards, from the highest to the lowest interest rate.

3) Look at the compounding schedule of each to see which card actually costs you more. The more frequent the compounding period is, the more expensive it is.

4) If needed, consider getting a loan to pay off your credit cards. This will help you switch to a lower, (simple) interest rate from the higher (compounding) ones.

5) If the lump sum amount and immediate schedule don't work for you, and you need more flexibility, get a credit line, and use that to consolidate your debt.

6) Remember the Rule of 72, divide your interest rate into 72 to see how long it will take you to double your credit card balance. Remember that your credit card balance is the bank's investment in you. That's how they look at it, *you'd better start seeing it that way too.*

7) Based on Rule of 72, now you know two things:

   a. It doesn't take extremely high returns to double your money in a reasonable amount of time, so there's no need to take undue risk.

   b. Time is of the essence here. The earlier you start, the more chance you'll have to double your investment, which can grow exponentially in time.

8) As a result of all of this, your summary of the summary is to immediately start paying off your debts, and get onto the path of saving.

6.4 — Create a Financial Plan

On **page 44**, after materialism's effects on our wallets, I listed the importance of having a budget and a plan for financial freedom. I'll give a quick recap of that here and then we'll dive into the subject of retirement planning.

You wouldn't try to erect a building, go on a long trip or cook for a large party without a blueprint. How can it be any different for your finances? Without a financial plan, hoping that it will all work out in the end, is no different than hoping your skyscraper will withstand an earthquake without the proper structural engineering. Even a visit to the hairdresser must always be scheduled on the calendar, yet, very few people have written financial plans. Is this important? Study after study has shown that people who write their goals down have a significantly higher probability of achieving them.

There's a long list of reasons for this phenomenon, and a few that I can anecdotally confirm are:

1) When you write your goals or ideas down, they get solidified, gain a physical presence on a piece of paper, and so become harder to ignore.

2) Most people remember the things they write down better.

3) Progress is easier to track with specific goals.

4) The accountability factor is wholly missing when your goals are not documented.

So, a good plan has to be objective, realistic, specific, and in writing. Here, Einstein's famous words are extremely valuable:

Everything should be made as simple as possible, but not any simpler

This also reminds me of the problem-solving principle of Occam's Razor. An English Franciscan friar named William of Ockham (1287-1347), made the argument that when there is more than one solution, the simplest is probably the right one. The more complicated your plan is, the more difficult it will be to execute, and the more prone to errors it will be. But you can't just throw some numbers on the wall hoping that they'll stick either. Finding the right balance, as always, is the key.

These are the six main areas of financial planning: retirement, insurance, investments, tax, education, and estate. I'll focus on retirement and investment planning. A Certified Financial Planner™ can be an important ally and advocate for you, playing the role of quarterback in this process. But you'd need an estate attorney, a tax professional, and an insurance agent to handle other specialized needs and execution, as these fall outside the realm of expertise of a financial planner.

6.4.1 — Retirement Planning

Most people traverse their span of working years hoping they can comfortably retire someday. Therefore, discussing retirement is helpful to pretty much everyone. In addition, the mechanics of retirement planning can be put to use in other areas of personal finance, because of important factors such as the time-value aspect of money, cash flow, inflation, and projected portfolio growth. You can also use this information for education planning, business income and expense projections, budgeting, and other related financial topics.

To begin, the questions you want to ask yourself are, using your best estimate:

1) How much money will you need at retirement?
2) When will you retire?

I ask these questions to my clients on an after-tax basis, in today's dollars, because your planning software will do the rest, but if you'd like to keep it simple, here's a tool which AARP provides on their website at this link: https://www.aarp.org/work/retirement-planning/retirement_ nest_egg_calculator/

This will help you find out how much you need to save today to reach your future retirement goals. Now, it's up to you to go ahead and start saving and investing toward that goal, and make it a reality.

A heads up or two about their site and other similar ones: make sure to expand each section, using the drop down menus, so that you can read through every area of interest. For instance, at AARP, to choose whether or not to include Social Security retirement benefits in your calculations, you need to expand the page. Otherwise, that information is hidden. On that note, many advisors will prepare their clients to set a target goal that excludes Social Security benefits. This is a matter of opinion.

If you prefer to do that, obviously the more the merrier at the time of retirement, and you'll be even better prepared with that added in. That being said, in my experience, for many people Social Security retirement benefits play a huge role, and without that income, the savings goal can become unrealistic. As a result, I do include it in most of my calculations. To set an unrealistic goal is as good as having no goal.

The estimator I shared with you above and many others you'll come across on the Internet may not factor in taxes, so it falls upon you to do that as well. At retirement, generally the expectation is that you'll be paying taxes in a lower bracket. You'll need to factor in your income tax rate as well. I don't want to throw out a number here without knowing the details that affect you, and so will leave it at this: save enough to prepare yourself for potential taxes as well. For more, consult your tax advisor, or look at the current effective (not marginal) tax rate on your tax return, and keep appraised of it.

For those who will not visit the AARP site, I've put together the following table, to provide a rough estimating guide as to how much you should save to meet your retirement goals. At the very least, this could be a starting point. The way to read this is: find your age across the top row, the income you desire (*including* Social Security on the far left column), and then find the approximate corresponding monthly savings needed. These figures are after-tax, so you'll need to save extra for taxes as well. (The assumed annual rate of return before retirement is 7%, and after retirement 4%, expected increase in income 3%, retirement age 67, retirement period 20 years, expected inflation 2.9%.) Social Security retirement benefits are computed based on income equal to the retirement income need showing in the far left column. Lastly, these figures are per person. If married, and the spouse needs an equal amount of income, these figures need to be doubled.

	25	30	35	40	45	50	55
$20,000	$247	$306	$385	$492	$649	$899	$1,367
$30,000	$433	$538	$676	$866	$1,142	$1,583	$2,410
$40,000	$619	$769	$967	$1,238	$1,634	$2,265	$3,450
$50,000	$805	$1,000	$1,258	$1,612	$2,127	$2,950	$4,493

So if you're 30, and you'd like to receive $40,000 a year retirement income at the age of 67, then you need to save $769/month today, assuming that you'll get a 7% return on your investments before retirement. If you wait another 5 years to age 35 to start, you'll need to save $967/month.

The important fact that this table shows, is that each 5-year period you wait to start saving, costs you progressively more. For instance, between the ages of 30 and 35, to receive the same $40,000 retirement income, one must increase their monthly savings by 26%. The same person, having waited until the age of 50 and 55, would need a savings increase of 52% to catch up. So, the cost of waiting is enormous and every little bit that you add to your savings in the earlier years helps.

For instance, for a 40 year old person, with a retirement income need of $50,000, if they had $50,000 in the bank, the required monthly savings would drop from $1,612 to $1,259, a 28% difference. How much does a 24-year-old need to save a year until the age of 40, a 16-year period, to have $50,000? With money market rates, less than $3,000 a year. Once the good habits are established, it's not that hard.

One between the ages of 30 and 45, looking to get $30,000 to $40,000 retirement income per person, including Social Security, you'll see that the required monthly savings amounts are anywhere between $538 and $1,634.

As previously mentioned, there are two ways to do this:

1) Have a budget and include also the amount of your ultimate savings goal in your budget.

2) Pay yourself first, then transfer your monthly goal to a savings vehicle and invest, *before* you start spending it, and make do with the remaining balance.

The first choice is the more ideal method and it leaves room for customization. But in my career, what I've seen is that it's extremely rare. The second option, paying yourself first, is the more practical way because it's simpler and automatic.

What if you'd like to buy a house, pay for kids' college expenses, and have a budget for traveling? You'd keep repeating the same process for each goal by asking the same questions:

How much do I need?

When do I need it?

How much do I need to save today to reach this goal in the future?

Add all your goals up, find the total amount needed, and pay yourself first for your goals. This is the essence of financial planning.

The good and bad news here is that everybody has a plan. They may not like it, they may not even be aware of it, but life has a built-in plan for all of us. According to this plan, we will get old, be forced to retire, eventually get sick, and die. That is the plan. There are costs associated with this plan that life has drafted for us. The question is, would you like to be in the driver's seat along the way? You, or randomness? Either way, a plan will be executed on your behalf. It's up to you to choose your path on this journey.

6.4.2 — Tax Advantages of Retirement Savings

Having helped affluent investors across the majority of my career, I can tell you that one trait stands out and differentiates each one from a person who just gets by, living paycheck to paycheck, and spending all or most of their income — that trait is their persistent search for savings opportunities.

When I worked at a major bank, I would notice that clients with higher balances would get better deals than an average client, such as zero monthly fees, or other service charges such as wires and checkbook order costs would be waived. A more average client, on the other hand, would pay for all of those services, and more. The person who needed the least would get all the help, and the person who needed the most, would get no help at all.

This was of course partly because the bank made enough money with the profits (margin) it generated from higher account balances, but also simply because affluent clients asked for deals regularly, whereas a "small customer" hadn't even thought about it. It's a different mindset. Affluent people are used to asking for deals, and they usually get them as a result — and for just simply speaking up.

There is one skill, I had to intentionally set my mind to learn. It's quite fascinating to see how easy it is once you get used to it, especially in business. The answer to the question, "Is there a discount you can apply?" or "Is there something you can do about this price?" is more often than not, "Yes, of course, let me see what I can do."

Another example of this phenomenon, is the attitude that affluent investors have regarding their financial dealings. They never leave money on the table regardless of how small it is, knowing that it all adds up. Conversely, it never ceases to amaze me, when I see a struggling investor being unmotivated to take advantage of ways to save money, as if their effort is not worth the price. I find this very ironic, indeed.

The tax savings of retirement planning demonstrates a perfect example of this. Within the IRS guidelines, your retirement savings in tax-qualified accounts, are deductible from federal taxable income. This means, depending on your tax bracket, one gets to save taxes on that amount, or in other words, gets an immediate guaranteed one-time return on their investment at their effective tax rate. Here's an example:

For an investor who is at 22% federal income tax bracket and 9.3% in California, (for a total of 31.3%), a $1,000 contribution to an Individual Retirement Account (IRA), can help save $313 from income taxes. In other words, it's like they have contributed $687, Uncle Sam gives $220, and the State of CA matches with $93. These dollars, if not contributed to an IRA, would otherwise have to be paid as income tax.

In 2020, the maximum IRA contribution limit is $6,000 for those younger than 50 and $7,000 for those 50 or above. That translates to a $1,878 savings in taxes for an investor younger than 50.

Even better, in another attractive retirement savings vehicle, the employer-sponsored retirement plan known as a 401k, the maximum allowed contribution of $19,000 for those under 50, (increased to $19,500 in 2020), offers a $5,947 savings if computed using the same annual tax brackets. For a couple, the computation gets more complicated of course, but to keep it simple, this would mean over $10,000 in tax savings, per year. In a 30-year career, with growth in investments and company matches, these dollars have a realistic chance to exceed a million dollars.

The tax advantages of retirement savings in qualified accounts do not end there. Your contributions' tax savings continue alongside gains in your investments. In a regular brokerage account into which you've invested with your after-tax dollars, you pay capital gains taxes for those gains that you've "realized," meaning you've sold and locked in your profit. In retirement accounts, taxes on these gains are "deferred," just like taxes on your income, to a future time when you withdraw funds. This allows your account to grow much faster, as your tax savings get invested and reinvested, over and over again.

A typical objection from a person who hasn't seen their account balances rise over time as a result of regular contributions and investment returns, goes something like this: "Yes, but I don't have $6,000 to set aside, I need that money!" or, "Yes, but you pay taxes when you retire anyway!" Here is where the rubber meets the road: you'll never hear these excuses from an affluent investor, or a person on their way to becoming wealthy. If you catch yourself having similar thoughts, please note that: to turn your finances around to stability, first you have to stop thinking with a scarcity mindset, meaning focusing on what you don't have, and start focusing on what you do have, and what you can do with it. Similarly, shift your attention away from what you can't do, and toward what you can do.

The technical answer to the latter objection, that is, paying taxes at retirement, is that for most people, the tax burden at retirement will drop, as a result of having less income to report for taxes. The answer to the first objection is simpler: do what you can, save what you can, because you have to start from somewhere. You must have heard the phrase; your hardest earned million dollars is your first one. It's mostly true. But it's also true that your hardest saved ten thousand dollars, is the first one as well. It's all relative. Because once your money starts working for you, there will come a point where your investment returns will improve your net worth faster than your savings rate. So, please stop looking for excuses and start saving for your life goals today. Every journey starts with the first step.

6.5 — Invest According to Your Plan

If you're still with me at this point, and on board with spending less, having a budget, and saving toward your goals, now you're ready to ask the next question: how to invest your hard-earned and saved dollars? The answer to this question is not only complex, but is also constantly changing. So it's important to become comfortable with change, uncertainty, and limitations. Keep in mind, that all of these qualities of investing are applicable to professionals as well as beginners.

One of the best bits of investment advice I've ever heard came from the legendary Warren Buffet, a.k.a. the Oracle of Omaha. His advice is not only critical for investing but also life in general (as usual):

The best thing you can do for yourself is to know your limitations

This advice may not necessarily help you make money, but it may certainly keep you from losing a lot. To refer back to Warren Buffet, he says that there are two rules of investing. The first rule is not to lose money. The second, not to forget rule number one. In that spirit, I'll begin with what not to do, and then direct you to the next more rational actions to take.

I should also add that, in the realm of investing, there are very few generally accepted strategies' as so many aspects of it are subjective. That's challenge number one. Challenge number two is that the generally accepted rules, such as buy low / sell high, are easier said than done. *Execution is all that matters.* In short, if you're an experienced investor, and disagree with what I have to say here, that's OK and normal. I'll always recommend what I believe to be the best methods based upon my experience, but depending on the situation, your disagreements can be more than valid.

First off, let me start by suggesting that you should *never get into a trade that you don't fully understand* — unless of course, you see it as a gamble or learning opportunity with a potential tuition fee that you're willing to pay in terms of losses. Start with the basics and believe me, investing is not like the Olympics. You don't get compensated for the complexity of your strategy. You are paid to take the right position in the market that will serve your financial goals. That's it. This incentive can easily be satisfied with a simple investment strategy and so there is no need to take undue risks. Your goal is to make the best calculated guesses from limitless possibilities, that will ultimately serve your interests in a future that is inherently unknowable.

The second "not to do" is this: do not have overly concentrated positions. In other words, do put your eggs in different baskets, because if you drop one basket, the others will survive. Here, try to find investments that react differently to market conditions. During the end of the NASDAQ bubble in 2000-2002, those who believed they were diversified enough by their having invested in different technology stocks, found out that their portfolios saw no added benefit of diversification because they were invested in highly correlated stocks — and together, they all went in the same direction: down. Even worse, they stayed down for 7 years, until the next bull cycle began in 2009.

In short, stay away from investments you don't understand, and resist the urge to have concentrated positions. If you avoid these two common, fundamental errors, then you'll also avoid irreparable damage to your portfolio.

Let's get on the same page regarding some terminology you'll encounter in this book, as well as in most other resources on this topic.

6.5.1 — Strategic vs Tactical

I want to begin by building a framework to help us avoid confusion as we move forward. The joke, that you'll get three answers from two financial advisors, is funny because it's so true. The subjective nature of investing means that there are always more than one or two ways to look at a thing, and potential misunderstandings hide in every corner.

I should also highlight to the reader that the meaning of terms used in this section is strictly limited to how each pertains to liquid portfolio investing. In other words, when you see a phrase, please consider its meaning within the realm of investing in stocks, bonds, cash, and alternatives.

Let's start by looking at strategy versus tactics:

A strategy is your longer-term plan that will serve your vision, while tactics are the changes you make as needed.

For instance, your strategic goal is to retire at 65 with 70% of your current income. Your strategy then would be to save and invest accordingly, based on historical returns, inflation, and tax assumptions. This strategy needs to be supported by an asset allocation that serves this vision, based on longer-term investment returns. This is why strategic asset allocation should be your starting point. I cannot emphasize enough the importance of this concept here:

The number one factor contributing to portfolio returns is the choice of asset allocation

What does asset allocation mean and why is it so important?

Imagine if you were to have bought a house in California back in 2009, and especially if it were closer to the coastline. What would be the probability of experiencing a financial loss from that purchase? In hindsight, we can say that probability was extremely low. Similarly, if you'd bought 100 stocks in the US around the same time, 10 years later, how many of them would we see had become worth less than their original cost? Unless you're extremely good at picking losing stocks, the answer is, very few of them, if any. Given this information, try to answer this:

In your opinion, which of these two decisions would be more critical if you were able to go back to 2009?

1) Should I buy a house now, or a stock now?

2) Which house or which stock should I buy?

The decision to buy or not to buy a house or stock is a lot more important, and your timing of this choice contributes to your investment returns much more significantly, when compared to the question of which house or stock to buy. There are many academic studies on this topic. Anywhere between 70-90% percent of investment returns depend on asset allocation choices. The most commonly-used asset classes are stocks, bonds, cash, commodities, real estate, foreign currency, and derivatives. Some would lump the last four into one, and say: stocks, bonds, cash, and alternatives. In summary, how much you invest, (in stocks, versus bonds, versus cash, or alternatives), is the single most important driver of investment returns, not the specific choices within each asset class. Let's see what Warren Buffett says about this:

A rising tide floats all boats, only when the tide goes out do you discover who's been swimming naked

Where does the choice of strategic vs. tactical come in to play? Strategic asset allocation determines the percentage of your investments that are allocated to stocks, bonds, cash, or alternatives. Tactical adjustments are the movements you make within this allocation, however often such changes are needed. Depending on your risk level, age, income, savings potential, and your investment objectives, your strategy will differ. As market conditions and your financial circumstances are a moving target, most likely you'll want to make tactical changes, with a shorter-term duration in mind as a result. But that doesn't necessarily mean that your strategic goals or allocation has to change as well.

Some will argue that your strategic asset allocation should never change, and since you can't time the market, you should not even aim for tactical moves. I, obviously disagree.

Depending on the business and economic cycle, some sectors or industries do better than others. Also, depending on the interest rate cycle, some fixed income securities will do better than others, based on their stated rates, risk, and maturity. So, there's no reason to blindly stay where you are with your investments, when the writing on the wall suggests that you do otherwise. *Tactical changes can be made both within the strategy, or can include minor changes to the strategy.* The key consideration is to keep the integrity of the strategic allocation in line with your goals.

Here's a demonstration of strategic positioning and tactical changes. Keep in mind that this is a loose template. Without knowing your specific circumstances and investor profile, any recommendations would be doing a disservice to you. Your life goals, income level, savings rate, and risk profiles have to be factored in while building a portfolio. The demonstration which follows on the next page will reinforce my point.

In this example, if you're between the ages of 20 and 30, your stock to bond, or risk-on (RON) to risk-off (ROFF) allocation could be somewhere from 60% stocks, 40% bonds, or to 80% / 20% respectively. As you age, your risk levels should drop all the way to the opposite side of this spectrum because you begin to have fewer years to make up for losses.

Age Brackets

	20-30	31-40	41-50	51-60	61+
20/80					X
30/70				X	X
40/60			X	X	X
50/50		X	X	X	
60/40	X	X	X		
70/30	X	X			
80/20	X				

Stock to Bond Ratio (Risk On/Risk Off)

There are three stock to bond levels for each age group for a variety of reasons. A 10-year spread in age is a big gap, and age is not the only risk factor, so the stock to bond ratios need to reflect that. These three levels can also be used differently depending on the where you are in the market cycle. The most logical thing to do, but also one of the hardest things to do, is to be greedy when everyone is fearful, and fearful when everyone is greedy.

At times, it could make sense to have a higher stock ratio when times are really bad, and to have a lower stock ratio when things are really good, as most likely, the end of the cycle is near. Buy when everyone is selling, right? Well, it's so much easier said than done during a crisis, but this could be a situation where tactical choices versus strategic positioning among asset classes, could be called for.

What is described in the paragraph above is a 'contrarian strategy,' and it may sound as if it's against the common principle of 'making trend your friend.' The idea is to keep the trend your friend only during the upward trend. You don't want for the trend to be your friend during a downturn (unless you are shorting). So then, what to do? Follow the trend until extreme levels are reached, then turn contrarian. Is it easy to do? Absolutely no, it's not. It is almost impossible to know you're in a bubble until after it's collapsed. As one of my previous bosses loved to remind us: no one said it was going to be easy. So, switching to a contrarian tactical position while actually still experiencing the positive trend, is one way to execute this, but it does require going against the trend for a period of time to capture the reversal, and it isn't for everyone.

You'll notice there are no zero percent stocks at the most conservative, and no zero percent bonds at the most aggressive ends of the risk spectrum. This is because studies show, that as a result of the negative correlation of stocks and bonds (usually), and the volatility in each of these asset classes, a 20% exposure to the opposite ends of the risk spectrum may provide added stability, and improve returns. This is a rare opportunity to have some free lunch. Usually, higher returns would require higher risk. Here, the investor gets a chance to lower their risk and increase their returns at the same time.

You'll also notice that there's no cash allocation here, and there's a reason for that. Everybody should have some cash put aside for emergencies. There's no hard and fast rule for this, because there's no easy way to quantify and form a generalized rule for how much to put aside for emergencies as a result of their random nature. But it wouldn't be unwise to put aside enough money to cover your monthly expenses for six months, in case of a job loss or illness. For a couple, some financial planners lower this to three months, as one of the spouse's income can provide the added buffer at times like these. I'd recommend to do what you're comfortable with, but your minimum amount to set aside in cash should be three months' worth of expenses, (and the maximum should be about a year's worth of expenses in cash).

These funds should not be invested in any vehicle other than a 3-month CD, or some type of money market funds. This will not only give you the needed peace of mind, knowing that you have money set aside for unforeseen events, but will also improve your investment returns for reasons that may not be immediately obvious.

Here's how it works:

If you invest all your funds including cash, during the first downturn, you may feel nervous and scared, so much so that you may do the exact thing an investor shouldn't be forced to do: sell at a loss when it can be avoided. But if you have some emergency cash on the side, you can tell yourself to relax and ride the downturn. There is no immediate need for these invested funds anyway, at least not in the short term, and this added calmness will help you make a better, cool-headed decision. This will improve your long term returns by not realizing every loss along the way.

These types of choices fall more into your strategic positioning. To get deeper into your tactical changes, I must describe the different asset classes, and the risks associated with each one of them, which brings us to the next section.

6.5.2 — Asset Classes and Risk

In the world of investments, when the term "asset class" is used, it usually means stocks, bonds, cash, and alternatives. Stocks, bonds, and cash are self-explanatory, but what are "alternatives?" In short, alternatives means every other asset class, including: real estate, precious metals, industrial, and agricultural commodities, foreign currencies, exotic cars, fine art, digital currency, and the list goes on.

In this section, we'll examine the risks, and look at what can happen when a desired outcome doesn't materialize. It turns out that portfolio losses have a larger impact on your investments than gains.

Here's an example:

If you invest $100 and gain 50%, your ending balance will be $150. That's great, but if you lose 50% next year, you won't be back even at your original $100, but rather you'll be at $75, or a 25% loss from your starting point. This is probably why Warren Buffet's first rule is, do not lose money.

Let's change the order of events. In this scenario, you lose 50% in year one, and end up with $50 from your $100 investment. Next year, you'll need to double your money, which is a 100% return to break even. A loss of 50% requires a 100% return to break even, (without factoring taxes and fees).

Because of the mathematics behind this relationship, limiting your losses should be your first goal rather than trying to hit the ball out of the park. This is precisely why I wanted to add this section to my book, because it's too crucial a point to leave out. This reminds me of a common phrase in sports: *offense wins games, defense wins championships.* Investing is a long tournament, and your defense needs to be tight, just as it must be in sports.

What are the asset classes and their risks?

A stock, also known as an equity, is an ownership stake in a company. There are two potential positive outcomes from an investment:

1) You can sell your investment for a price higher than your initial cost, which is capital appreciation or growth.

2) You can receive income from it.

When you buy a stock, depending on your selection, you make yourself available to one or both of these favorable outcomes. For most investors, the main goal of buying stocks is capital appreciation, and the income in the form of dividends is the cream on top. In most cases, those who seek income would typically invest in bonds, because of the fixed income they provide. There are, like in anything else, exceptions of course. Some investors do use high dividend-paying stocks as part, and sometimes a large part of their strategy, which is to focus on investments which generate income. This is becoming more and more common in today's low-interest-rate-environment, and there's currently no end in sight.

The potential gain from a stock investment is easy to guess. As a shareholder in a company, if the company does well, and improves its earnings, its market share, its revenues, and its profits, then the market value of your investment will go up accordingly, hence the name: capital appreciation.

But what is the risk? You can lose all of your money, 100% of it. If the company goes bankrupt, you may not be able to recover any of it. This is one of many reasons why most advisors are advocates of diversification. If you were to decide to invest all your money in one stock, you'd want to keep in mind that there are many instances in history when large and seemingly unshakable companies simply went belly-up, (just like Enron and Lehman Brothers). In times like those your investments would go down with the ship. But consider this, if you were to invest 1% in 100 stocks, in that scenario, the most you could lose would be 1%.

The usual risk and return relationship of a stock is (assumed to have been) set by the market participants, and usually the higher the potential growth of a company, the higher its share price relative to its earnings. These are typically called growth companies. But also, a disappointment in earnings can have a larger effect on such a stock's share value.

Conversely, value stocks have typically a lower market value relative to their earnings. As a result, they are less prone to earnings disappointments and price fluctuations. Depending on the business cycle, sometimes growth company shares will do better, sometimes value shares will perform better. And this is where your tactical allocation can come in handy.

Toward the beginning of a business cycle, growth company shares may be favored by the market, as they expose their investors to perceived growth. At such times, also interest rates may creep up, but this doesn't affect dividend valuations of growth stocks, as many don't pay dividends.

Conversely, many value stocks pay higher dividends, and their valuation is adversely impacted by rising interest rates. Toward the end of a business cycle, this relationship may reverse. It doesn't always though, as almost nothing about investments repeat themselves perfectly. So you'll want to make tactical changes accordingly.

Examples of growth companies are technology firms and producers of discretionary consumer products. Examples of value companies are utility and telecommunications companies, along with those which sell household staples.

So, here's a perfect example of tactical changes within a strategy. You can keep your stock to bond ratio static, but make changes to the sectors you're invested in, and that would be a tactical trade within a strategic allocation.

Our next asset class, bonds, or fixed-income securities are designed to provide, you guessed it, a fixed income, which is usually called the yield or the interest rate. This doesn't mean though, that their market values won't fluctuate. As a result, this is probably one of the most confusing parts of this type of investing. Many people are shocked when they lose money on their bonds in the open market, and as a result, they get hurt by what they perceived would be a safe investment.

A bond promises to pay interest at certain intervals. The higher the credit rating and quality of the issuer, the higher the chance this promise will be kept. So far so good. But if the investor wants to sell the bond in the open market, for one reason or another, before the bond matures, they may not get their full investment back. That is to say it may lose some of its principal. How is that possible? You thought bonds were safe investments? You are not alone. Here is how it works.

Fixed income market values and current interest rates move in opposite direction (inverse correlation)

If you buy a bond that pays 5%, and the FED decides to increase interest rates after, there may be newly issued bonds in the open market similar to your holding, and they may pay 6%. In that case, your existing bond will lose its attraction to the next potential buyer and drop in value. The most common analogy I give to my clients is that of a car. Let's say that you have a car that goes 30 miles per gallon. Tomorrow, a new car at the same price gets on the market that goes 35 miles a gallon. Naturally, all things being equal, the value of your car will drop. Similar mechanics are at play with fixed-income investments. The income of the bond is fixed, similar to the miles per gallon covered. To reflect the effect of the newly issued offerings, the only adjustment the buyer can make is to the bond's market value.

So then, when you invest in bonds, the risk of not getting your principal back isn't only confined to what's happening with regard to the credit quality of the issuer, but also the current market conditions if you should need to have your investment proceeds before the maturity date. Naturally, a compromise exists here as well. The lower the credit rating of a bond, the higher the interest rate it usually offers to attract investors for taking on this risk. The good news for the investors in this particular type of vehicle, (in most cases), is that mathematically, the higher the yield of a bond, the lower its sensitivity to changes in interest rates. Here is an example:

You hold two bonds, Bond A and Bond B having 2% and 8% interest rates respectively. If the FED decides to raise interest rates by 1%, this change is 50% [1/2] of the interest rate and income to be paid out by Bond A, and 12.5% [1/8] that of Bond B. As a result, in comparison, the market value of Bond A will be impacted significantly more.

This relationship is not a 4:1 ratio, meaning the values of these bonds will not change at a rate equal to the interest rate differential, because there are other factors such as maturity date, (the longer that date, the more sensitive the bond is), and quality ratings, both of which will affect the supply and demand relationship of the bond.

So, armed with this information, let's look at a scenario, which shows that holding stocks and bonds can be beneficial as a diversification tool. Imagine a time when the economy is doing well, jobs are plentiful, incomes are rising, and so is the demand for goods and services. In a typical cycle, this would create an imbalance in supply and demand, and so prices would go up and create inflation. Rising inflation would push interest rates higher, and as explained above, this would put pressure on the fixed income portion of an investment portfolio. But at the same time, in such a scenario, stock investments would likely do well, and that portion of an investment portfolio would grow. As a result, depending on your allocation percentages, you might be losing on your bonds' market value, but could be making up for those losses (or could be doing even better), because of the growth of your stocks. If you reverse this scenario: the economy is slowing down and the central bank is lowering interest rates to provide a boost, this time bond values would likely go up, and stocks would likely have a hard time holding their own.

As you can see, diversification can lower your risk and can potentially improve your long term returns, which is the only free lunch that I am aware of. In fact, here is a demonstration for those of you with analytical minds. Pick two exchange-traded funds (ETF), one is an aggregate bond fund, the other an S&P 500 stock with symbols AGG and SPY respectively. Look at their total returns. Here, UP reflects better rate of return *compared to the previous year*, DOWN reflects worse:

	AGG	SPY
2008	7.02%	-35.86%
2009	2.99% DOWN	22.22% UP
2010	5.69% UP	12.96% DOWN
2011	7.37% UP	0.8% DOWN
2012	3.16% DOWN	14.13% UP
2013	-1.91% DOWN	28.74% UP

I picked a six-year period where I could easily find some dramatic market movements. As you can see, in all six years, when the bond returns improved from the previous year's, stock returns then deteriorated, and vice versa.

So, if you have a longer time horizon, say more than 10 years, taking risk may be compensated with higher stock returns. But if your time horizon is less than 10 years, diversification is even more important, as one bad year can wipe out all your gains. In such a case, bonds will likely provide the needed cushion for the hard landing, as we saw in the case of 2008.

Most people focus on the return side of the investment equation and don't ask enough questions about the risks involved. In the world of personal finance, the risk is the likelihood of a scenario in which you lose money. The larger the set of probable outcomes, the larger the risk exposure you have. The trick is though, as seen in the bell curve on the next page, your risk to the downside, (left side of the curve), is in most cases, directly proportional to your risk to the upside, (right side of the curve). In other words, investments with the potential to go up 10%, also have the potential to go down 10%.

Now, of course, not all investment outcomes display a perfect bell curve shaped outcome, (although it's true that a few do), and it's surely impossible to know what it will look like beforehand. But without the guidance of hindsight, I suggest you assume that your upside potential is always matched with an equal downside.

Even the safest investments, like bank CDs, have their own set of risks, such as losing purchasing power against inflation. If you live in a time of rising inflation, the safest way to lose money would be investing in a CD, because you'd be locking your money into a low interest-bearing instrument that barely fights inflationary pressures.

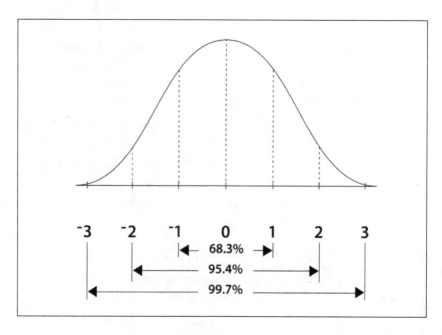

Bonds are nothing more than a lending instrument and default risk is a significant factor. What if you don't get your principal back? This is why some investors prefer to focus on the return "of" their capital, as opposed to "on" their capital, as with whatever are their specific circumstances, it's OK to play safe and target the preservation of capital first.

In the particular case of risks involved with "shorting stocks," you have unlimited loss potential in this trade, as stocks have no appreciation limit to the upside, and you'd be taking the opposite position with regard to this potential.

In derivatives such as options/futures, you could find yourself at a 100% loss simply by seeing your contracts expire, or, depending on your contract, similar to shorting stocks, your loss can be unlimited.

Lastly, some feel that real estate is a safe investment, mainly because they fail to take into consideration risks such as floods, earthquakes, fires and other disasters, as well as changes that could come to pass in policies, regulations, maintenance costs, random sustained vacancies, or as in 2008, a market correction. A fresh update to this could well be in the making, with respect to the potential lasting effects of Coronavirus on, for instance, commercial real estate. There's a reasonable chance that the need for office space may be significantly diminished, with video conferencing practices proving to be so useful during this era of tele-working from home while "Shelter in Place."

In short, each investment has its own unique or common risks which should be understood well before you commit your hard-earned, and even harder-saved dollars. I've lived through the bursting of three major bubbles: NASDAQ, sub-prime mortgage, and now a Covid-19-induced drop in prices. I've seen people lose their entire life savings, or their house over the downturn of a stock or two. Financially, or maybe even emotionally, bouncing back from an experience like that is extremely difficult. So, it's very wise to invest with a strategy that takes into equal consideration the upside potential — and the downside.

6.5.3 — How to Choose a Financial Advisor

At the beginning of your saving and investing journey, you may not feel a pressing need to work with a financial services professional, or you may feel you have the time to educate yourself on the prescribed topics. But there will likely come a time where you'll find it difficult to stay current with all the tax laws, investment opportunities, and insurance products, that will be affecting you — or you'd simply like to have a financial plan. This is likely to happen as you start a family, or if you find you've diversified your investments into areas in which you're just not as competent, or you simply run out of time. In those instances, you'll look for a financial advisor, and I'd like to provide some guidelines for selecting the right one for you.

Let's start with the fiduciary capacity. In the aftermath of the 2008 Global Financial Crisis, triggered by the US sub-prime mortgage loan debacle, it became clear that many large financial institutions, such as insurance firms, banks, lenders, investment firms, and credit rating agencies, may not always have their clients' best interests at heart.

Most notably, investment firms could, (and did), see fit to sell stocks from their own accounts, while their brokers were busy buying the very same stocks for their clients' accounts.

Rating agencies were giving quality credit ratings to bundles of low-quality bonds, not fully understanding and /or disclosing the risks these posed to investors.

Insurance firms were insuring these products, too, without really understanding their risks, and the entire doomed system continued to be perpetuated by salespeople in high rises, who were collecting fat commission checks without considering the consequences of promoting these flawed products — never imagining that they could deteriorate as disastrously, as they eventually would.

As a result, once the brakes were too hot to function properly, driving off the runaway ramp was the only remaining option, as none of the existing checks and balances were able to stop the doomed vehicle.

As a result of this experience, one concept that had existed already for a long time, suddenly became very popular: that of the fiduciary. A fiduciary, by definition, is a professional who is legally and ethically bound to do what is in their client's best interest, not what is in their best interest. As a result, a fiduciary cannot do the things the large brokers did in 2008, such as selling unfavorable stocks to benefit their accounts while having their clients buy the very same ones to benefit only themselves. A fiduciary may not put themselves in any situation where a potential conflict of interest exists.

So, naturally I believe that if you ever decide to work with a financial advisor, you should seek one that is a fiduciary. How do you know if the person is a fiduciary? You can simply ask them, of course, but you'll find that if they work at a large firm, which is selling you financial products, they most likely will not be a fiduciary. (In fact it's not allowed in many cases, as the act of selling financial products such as mutual funds or insurance, by default creates a conflict of interest because of the compensation for the transaction. If I advise you to buy mutual fund A over mutual fund B, and if I get rewarded with a larger commission check for selling mutual fund A, what do you think I'd be likely to do? Are there firms out there whose agents can and do sell mutual funds without getting a commission check, and therefore can be fiduciaries? Yes, but they are very much the exceptions to the rule, not the norm.

As a result, most fiduciaries are independent and work on a fee-only schedule. This means, they don't sell products and instead, they work on a fixed hourly, or asset-based, fee schedule. By doing this, what's good for the client, is also good for the advisor. If the client's assets grow, so will the income going to the advisor — which is an example of an asset-based fee structure. Consequently, you don't have to do anything special to motivate your advisor to grow your assets, as you sit on the same side of the table.

People face different financial circumstances. A real estate investor who buys/sells investment property will be better off working with a professional who has experience in overseeing such transactions. Whether a family of four with two kids looking to plan for education and retirement expenses, a tech executive with stock options in their compensation package, or an investor with dual citizenship, each will have very specific needs. So, you should work with an advisor who has experience relevant to your situation, not just any experience.

In short, look for an advisor who fits your needs and don't be shy to ask them probing questions about the types of experience they've had. For instance: "How many clients have you worked with in the past few years, who have had conditions similar to mine?"

The financial planning aspects of personal finance, including: taxes, insurance, estate, and education planning, are so complex and crucial, that ignoring them could easily harm you over time. So, a financial advisor with financial planning knowledge can benefit you greatly. One way to make sure of that is to search for a professional who has earned the Certified Financial Planner™ designation. While not legally required, this title is very beneficial as it's bestowed by an industry-regulated organization, which ensures that the education and experience requirements of it's financial planners are not only met, but are also kept current.

Anyone who gives investment advice to the public for compensation needs to be licensed and registered with the authorities. This is a complicated topic as there are many different licenses and regulating bodies. So, I won't get into the details, but at the very least, you should raise the question to your prospective advisor, "What licenses do you hold, and where can I obtain a record of them?" If you don't get a satisfactory answer, do some research on the SEC's public website, at: BrokerCheck.Finra.Org. There you can seek such details for yourself, and verify not only that the financial advisors you're vetting are licensed, but also that they have a pristine public record.

Lastly and most importantly, have a heart-to-heart talk with the candidate, to make sure that your values and needs align with theirs. For instance, if you're sensitive to topics around Socially Responsible Investing and your advisor isn't, this difference carries the potential to cause a fair amount of friction between the two of you going forward. Additionally, your financial advisor's comfort with risk shouldn't be too different from yours. If you want to take extreme risks and get into speculative investments, your advisor needs to be able to feel comfortable with that. These are just a few examples of how your values need to align with your advisor's. Have a full and healthy data-gathering meeting to find these things out.

Not everyone needs a financial advisor, that's true. But if you think you'd benefit from their expertise, doing your homework before you meet will help you greatly. Consider this check list. I hope it helps and good luck with your search.

It's preferable for your advisor to be:

- Independent
- A fiduciary
- Fee-only
- Licensed and registered
- A financial planner as well
- Experienced in relevant areas
- To be known to have a proven track record

7
How Mindful Awareness Can
Improve Your Finances

Mindful awareness or mindfulness can help improve every aspect of our lives, but specific to our purpose, it can help one stop overspending, and it can gently contribute when it's time to save and invest — *it has the power to be a most useful ally.*

What is mindful awareness? This concept was widely known for many thousands of years in the East, and it's become mainstream in the West across the past 50 years or so. Mindfulness as a technique and practice is used to improve mental and physical health in the military, academia, the private sector, and in healthcare. In short, being mindful means to be present and objectively aware of the internal and external conditions of the current moment. It's being able to remove the thick smoke layer of obscured understanding, to reveal instead what is. It helps us see things free of preconceived notions, judgments, prejudices, and assumptions, and frees us from the entanglements of the past and the future.

If you're looking for one skill that will have the biggest impact on improving your life, this one will help you see things as they are, by way of a simple assessment of facts. You'll benefit from having a firm grip on reality. Otherwise, if you aren't able to draw accurate conclusions, you'll experience a detriment to your life in general, and certainly also, a drag upon your ability to take control of your finances. Mindful awareness is a state of mind that has the power to serve these qualities to you on a silver plate.

For the most part, we're the product of our environment. Separated twins who've lived in different conditions are striking proof of this. Despite their identical genetics, they may display very different personalities, values, and life choices. Even the language you speak can work as an overriding mechanism in setting the algorithm of your thought processes. Similarly, our financial conditions have also been directly influenced by external factors, such as consumer trends, educational systems, laws, regulations, and economic conditions.

To be truly free of these influences, we first have to identify the conditions of our surroundings and analyze how they may be affecting our actions. Otherwise, if we're on autopilot, we risk becoming nothing more than willing slaves as we continue to simply react to our conditions. This, in and of itself wouldn't be the worst thing if we had full control of these externals, but the exact opposite is the reality for most of us. Once we understand this, then we can take control of our choices and behavior patterns, which will ultimately determine our financial status.

Don Draper of Mad Men, has a great line about this:

"People will show you who they are, but we ignore it because we want them to be who we want them to be."

Let's remember the biochemistry of our thoughts, feelings, and emotions, and how they impact our behaviors, which I've discussed in detail in earlier chapters. If you're not aware of these factors, you'll run the risk of being controlled by them. Mindfulness will help you flip this equation, help you control these variables, and in turn, empower you to control your life.

Mindfulness can help you be aware of your thoughts, feelings and emotions with a lot more clarity, and instead of reaching for unsustainable, unhealthy solutions due to some deficiency of neurotransmitters and hormones discussed in Chapter 4, you can develop healthy habits that will help you resist the destructive band-aid of knee-jerk consumption.

Let's return to the question on the cover of this book: Why can't most people save money for their future needs? We've covered so many relevant topics, we may now be able to answer this:

Here is the list of reasons most of us can't save money:

1) Because we're constantly bombarded by masterfully-crafted messaging telling us to buy this or that extra special product, so we can enjoy a purchase-induced illusion of happiness for a fleeting period of time.

2) We live in a materialistic, capitalistic system, which doesn't prioritize our well being, but instead favors the growth of capital, hence the name "capital-ism." In the US, we're rarely, if ever, encouraged to save and invest, but only to consume.

3) Social and peer pressures from "the system" have pushed us to the edge of bankruptcy, as we tried to keep up with those around us. We're made to believe our self-worth is equal to our perceived net worth.

4) In the West, life evolves in a binary way as a result of duality being the overriding principle. We're trained to see ourselves as either successful (based upon our cars, houses, technology, gadgets, and clothes), or we're not. There's no in-between.

5) After survival, our strongest needs have to do with feeling validated and connected. In modern life, we can find ourselves chasing these feelings by buying things. Anything to fit in, from attending vacations to giving gifts.

6) The algorithm of our brain prioritizes feelings and emotions, and puts reason in last place. It takes a conscious effort to move our decision making to the more sophisticated region of our brain, the neocortex, where reason and logic reside. Those who fail to put in this effort, become reactionary organisms being pushed around by their feelings and emotions.

7) Because of the way our minds work, our thoughts are full of biases and fallacies that most of us aren't even aware of. But the corporations attempting to sell their wares to us, are acutely aware of the way our minds work. We have what I've come to call the Lazy Option Bias, which means that taking the easy route is often at the center of our decisions.

8) The addictive nature of shopping and spending money makes it simply very difficult to resist.

9) Nowhere in our education system are we taught the rules that govern personal finance. Unfortunately, for many people it's the reality of making multiple costly mistakes which leads people to want to learn about it.

Try to imagine which one tool in the shed possesses the best chance of overcoming each of these causes. Did you guess? It's mindfulness. When you become aware, you'll be less likely to go with the flow, and instead, you'll become motivated to take charge. The only way to navigate the challenges presented above is to live consciously. Mindfulness can be your GPS on that journey.

It can keep you from being reactionary and living on auto-pilot, responding to every stimulus served to you by corporations trying to control your attention. It frees you up to do what you want to do with your energy, time, and money — not what others want you to do with it.

By now, you either like living in a capitalistic society and you resent my comments, or you're a critic anyway, and have a few negative thoughts of your own about it. In either case, I propose, since you can't change your environment overnight, that you try to see it for what it is, and use it to your advantage. Turn the equation upside down and start looking for ways to improve your finances immediately. *You're not going to wake up to a revolution tomorrow, but you can start a revolution within yourself today.*

You can do that by being mindful of your true needs and how to best go about satisfying them. If you only buy what you truly need, and not more, you'll begin to see that there are dollars in your account at the end of the month. In short, your resolve to keep your money, has to be stronger than that of the corporations, who are so intent on taking it from you.

How do you improve mindful awareness? By improving your ability to concentrate and hold your concentration. Since I've mentioned meditation repeatedly, which is the ultimate cure for this problem, here is another technique: focus only on the task at hand 100%, and don't allow distractions. Start paying attention to where your mind wanders, and bring it back to your task. Like any other muscle, these skills which will help you improve your focus, will become stronger with practice.

Mindfulness is giving your full attention to the person in front of you, the food you're eating, the movie you're watching, or the task at hand. You will go where your attention is. A practice of Mindfulness will help you to focus your attention where you want it to be, and as a result, you will become who you want to be — and ultimately, you'll be able to create your own reality.

8

Philosophers Chime in on Happiness

If the number one reason that people overspend, beyond taking care of essential needs, is the quest for happiness, then it's the life-affirming activities which bring happiness and hold the key to relieving the stress that has led so many to overspend. In that spirit, I'll introduce you now to some time-tested ideas for finding happiness through natural means, as presented by some of the all-time great thinkers. These ideas will reinforce how a person can live a virtuous life without giving in to self-destructive habits.

Implementing the wisdom of sages is a life-long journey that requires years of practice, discipline, soul searching, and self-healing. Admittedly, it's also easier said than done. But if you set the expectation that your progress will be the one-step-at-a-time sort, and not immediate perfection, in time you'll go much further than you initially imagined.

Before I roll out the proverbial red carpet for the series of philosophers whose ideas I'll be presenting here, I'd like to highlight again, my favorite deep thinker on the topic, Epicurus. He is by far, the philosopher who dedicated the most constructive collection of thoughts and ideas to the intersection of money and happiness, which is ultimately the topic of this book. After years of contemplation, he concluded that happiness is dependent on a combination of three essential things: freedom, reflection, and friends.

Please do go back to the first chapter and revisit his ideas before you consider this book closed. If you're like me, and it takes multiple times to internalize new information, I highly recommend that you periodically review this content.

While reading the following sections, the key is to resist the common mistake of giving a quick dismissal. Often when we approach new philosophies or ideas, our natural instinct is to look for reasons to reject them. This could be due to some of the biases we're now aware of, or it could even be due to a simple survival algorithm, which scans for danger and looks for potential trouble. This is not necessarily a bad thing, but when learning new material, it can be counterproductive. Instead, if you look for what does resonate about the content, it may lead to a much more productive and beneficial read.

With that said, let's see what the philosophers have to say about happiness, the good life, and money. Here we go.

8.1 — Aristotle

We happen to be living in the most hedonistic period in human history. As a result, the majority of people are conditioned to seek happiness in bodily pleasures and material wealth. Aristotle (384-322 BC), the philosopher who's had arguably the biggest influence on Western Civilization's ideas regarding logic and thought processes, teaches us that these indulgent habits ultimately do not bring happiness. But instead, that it's the search for wisdom, and the practice of virtue, which can.

In his book *Nicomachean Ethics*, Aristotle[33] put forward the idea that the "highest good" for humans, can only be achieved through activities that make us human, and which ones in particular, separate us from animals. There's nothing unique to us as humans when it comes to our bodily pleasures such as having sex, eating, sleeping, resting, and physical play. In Aristotle's words, these are activities and pleasures even cattle in a meadow strive to maximize.

What separates us from other animals, is our ability to reason and contemplate. Additionally, we actually have an appetite for intellectual endeavors. According to Aristotle, these are the sort of activities we should spend most of our time pursuing, (learning, reading, writing, and expanding our skills).

Once you've embarked upon the path of seeking wisdom and virtue, eventually this won't be enough. And it shouldn't be. It's equally important to embody this knowledge and to learn to live within its parameters. Our personality, character, behavior patterns, and daily habits should reflect the information we uncover. In other words, we should practice what we preach.

A person who has internalized such favorable traits would naturally exhibit courage, generosity, and temperance, the combination of which, St. Thomas called, "a disposition of the mind which binds the passions."

If there's one word which does a fair job of summarizing the most important lessons in life, it would be "balance." According to Aristotle, the virtues of the intellect mentioned above, can help us strike a balance between not having enough and having too much. Just as what we call poison, is often only an overdose of medicine. Though essentially the same chemical, what determines whether the result will be healing or damaging, is the amount ingested.

According to Aristotle, a life in search of knowledge, lived in virtue and wisdom, in a balance struck by avoiding extremes, is a life he characterized by the word "Eudaimonia," meaning: happiness, wellbeing, or blessedness.

We like to think that we're in full control of our choices, and for some possessing a strong will and discipline, this could well be true. But for most others, it's unfortunately an illusion. As a result, being born in precisely the right environment is critical to achieving Eudaimonia. Aristotle advocated that it's the *State's role to provide an environment which is conducive to its citizens being able to attain happiness and a good life.*

The set of institutions which would help or hurt us on our path to this virtuous life, are the political and governmental systems we're born into. According to Aristotle, a *happy person is one who derives pleasure from good behavior.* They know the right thing to do, and they choose to do it. These people would encounter no conflict between theory and practice. Taking it a step further, this virtuous person, then, is who the State should support and help to develop through the right kind of education, and just laws. To be fair, the State would also provide the social programs needed to uplift the less fortunate. *With this idea, he became the first known philosopher to call on the role of the social state to help achieve Eudaimonia — a state of happiness.*

In summary, Aristotle taught us to focus on self-education, learning, and virtuous behavior. That is to say, that a person should aspire to do what's right and take pleasure in it, rather than seek material wealth, validation, and bodily pleasures.

8.2 — Socrates (469 - 399 BC)

Socrates, whom many consider to be the father of philosophy, contended that human happiness could be achieved through effort and the right mindset. Whereas many thinkers took sides on the issue of whether to rely on rationalism or spirituality in search of a good life, Socrates[34] advocated to find a balance between the two. He believed that following both reason *and* our soul's direction would provide the correct path to happiness. *Without constantly looking for a way to connect the dots, we would feel lost in randomness.*

In that spirit, he said:

The unexamined life is not worth living

Mostly barefoot,unbathed, and wearing old and dirty clothes, he famously walked around hassling the citizens of Athens with his annoying questions. Everything we know about him, we know through his pupils, the most famous of whom was Plato.

Socrates taught his adherents to question everything, reject all dogma, and challenge even facts as they're known. And so he believed, the best practice for uncovering truth, was to *ask the right questions.* This practice today is known as the Socratic Method, though in his day it agitated the wrong people, and cost him his life. He was accused of "corrupting the youth," and was sentenced to a death by poison. His most famous words are:

All I know is that I know nothing

Socrates was given plenty of time to prepare for a defense, that could also have been used to arrange an escape — which his pupils begged him to do. But he chose instead to face his challengers, certain that his death sentence was inevitable. In doing so, he still inspires us today, that we should stand up for our beliefs.

In the 1787 painting by Jacques-Louis David, The Death of Socrates, he's depicted in his bed, most likely engaging in a discussion on this very topic.

In his dialogue: *Euthydemus,* Plato quotes Socrates giving the first philosophical argument having to do with happiness. Socrates puts forth that happiness is the ultimate goal of all our actions, and that while *it could not be attained by external factors themselves, it could instead be attained only by utilizing those factors in the pursuit of happiness.* Perhaps something of an early voicing of the adage: happiness is to be found not in the destination, but the journey.

For instance, owning a pen in and of itself won't make us happy, but crafting handwritten notes to friends and family with that pen, can make happiness to us as well as the recipients of our letters. So it's our level of wisdom, which determines how our external behaviors will turn into practice, and spur us onward toward our goal of attaining happiness. In this, it's our capacity for logical thinking, rather than the blind following of desires, which is the key.

In the *Symposium* dialogue, Socrates argues that chasing our desires can only bring us happiness for a short amount of time, because there's no end to our desires, and so no end to the chasing. We need to train ourselves to take enjoyment in simpler things, and intentionally satisfy our souls with beauty, art, music, friendships, and love.

One cannot be happy without being virtuous, and conversely one will not be virtuous forever, if can't sustain their happiness by way of the virtue. Socrates had this idea in common with Aristotle, and with all the philosophers in the next section. *We can only find lasting happiness and live a good life, if we find it in the act of being virtuous.*

The secret of happiness is not found in seeking more, but in developing the capacity to enjoy having less

8.3 — Buddha

His real name was Siddhartha Gautama, a prince, whose date of birth and death remain uncertain, but are believed to be between the 6th and 5th century BC. His journey to enlightenment became a religion with more than half a billion followers. Here, I'll present him as a philosopher, not a prophet or religious leader. This preference doesn't stem from a lack of respect, but rather will keep us from potentially going off-topic. The teachings of Buddha are extremely rich and have many branches. As a result, I'll focus on his insights in answer to questions devoted to finding happiness and living a good life.

Buddhism's first noble truth dives right into the debate:

There is suffering

The main premise of this book is to show that in modern life, overconsumption is being used unsuccessfully to relieve this suffering, and even worse, this puts people in difficult financial circumstances, which creates even more suffering.

Buddha's second noble truth:

There is the origin of suffering

Buddha's third noble truth:

There is the cessation of suffering

And there is hope in this, the fourth and last noble truth:

There is a path to the cessation of suffering

The question is how?

That's what Buddha spent his whole life seeking to answer. First, in his methodology, he identified what caused suffering, and in the absence of those traps, a person would be left in happiness. This is an interesting way to look at the problem. According to Buddha, our natural state is to be happy, so there is no reason to seek it. The path to happiness then is in turning away from that which keeps us from being in our natural state. Rumi gave a similar message with regard to finding love. He said, love is everywhere, you need to remove the barriers you've created to it, and it will enter your life naturally. So then, the question is, what causes suffering?

The attachment to external conditions, all of which are in a constant state of change, and creating continuous stress and disappointment, is one of the sources of suffering. Some cling to youth, spending thousands of dollars to willingly undergo surgeries in an attempt to look younger. Most of us get attached to the people around us or to our pets, and when they inevitably pass, we're left with pain and suffering. Some can't let go of money, even for much-needed improvements, and live in difficult conditions as a result.

According to Buddha, the solution is to fully enjoy these things: our youth, friends and family, career, possessions, in the present moment. If we make the most of our years of youth at the right time, aging would not be resisted but accepted with grace. If we spend enough quality time with our loved ones in life, then their departure at death might not feel so devastating. Change is the only constant, and we have no control over our external conditions. We can only control our inner reality, which is our thoughts and feelings. This is what Buddha taught. What we need to cultivate is a disciplined mind that refrains from attachments. In doing so we become able to deeply appreciate what's happening in the current moment, and we can know and accept that, whatever it is, it won't last.

The second cause of suffering is our desires. There are certain teachers who interpret this to mean that ALL desires are destructive and need to be avoided. I promised not to share any knowledge which I've not personally experienced myself, so this total avoidance is not a view I'm on board with. The way I interpret this idea, is that we should train our minds to have the right kind of desires, and be able to avoid the destructive ones. For example, it's understandable to wish for good health or to desire the wellbeing of our loved ones, but it's not healthy to desire excessive power or money beyond what's necessary to living a good and dignified life. So we need to be able to differentiate constructive desires, from destructive ones.

Unfortunately, the challenge doesn't end there. We should also train our minds *not to be attached to our desires*, regardless of how constructive they may seem. For instance, one day, we'll all come near to the end of our lives, and quickly or slowly, we may see a deterioration of our good health. If we're attached to our constructive desire for good health, our last years may pass in suffering. But if we know when to let go and recognize this change as part of a natural process, we can see it as just another step in our lives, and not suffer over the loss.

So in short, according to Buddha, our attachments and desires are the most common roadblocks to our natural state of happiness. So the key to cessation of suffering is wherever we can dismantle these hurdles.

During his journey, Buddha tried many different methods in his search for enlightenment. There were many schools of thought, traditions, and methods practiced by many different gurus during His time, though many involved overcoming or transcending the ego. These ways included exposing the seeker to extreme hunger, lack of sleep, and in general, pushing the body to its limits. Buddha found that these methods failed to result in his goal of reaching Nirvana.

So instead, he advocated what he termed the Middle Way, which is to find a balance in one's actions. Eating enough to live a healthy life, not more, not less. Working the mind and the body to avoid atrophy and stay sharp, but also to rest and learn to be in stillness. This idea of the Middle Way, can also help us with our desires. For instance, it's healthy of course, to desire one's spouse in any strong relationship, but too much can turn into addiction or obsession, and ultimately this can become destructive. Conversely, the lack of desire can cause the erosion of a marriage over time. So balance is the answer.

Another good example can be seen in our current approach to the would-be stewardship of our environment. Regardless of scientific evidence which shows how our choices impact our total carbon footprint and the advancement of global climate change, our obsession with money, physical desires, and power, has overpowered our logic. We've lost the balance that once existed between producing for necessary consumption, and responsible and sustainable environmental impact. In the near future, humanity will face the consequences of this imbalance, in the form of ecological damage leading to life-threatening conditions.

Buddha also talks about the importance of enjoying good company. There's a theory that we become an averaging of the 5 people closest to us. I see that there's some truth to this. If you surround yourself with people who are always looking for ways to improve themselves, you will inevitably be affected by this. I, for example, have greatly benefited from living in Northern California, as part of the social life here includes being physically active outdoors, which is good for the mind, body, and soul. Repeating the same behavior patterns over a lifetime does make its changes to you in a certain way. It's critical to surround yourself with people who also look for ways to live a balanced and healthy lives. Buddha says it's better to be alone than spend any time with a fool.

Lastly, Buddhism teaches us that we're not our thoughts, feelings, and emotions — which is one of the main points of this book. At first this may seem odd and difficult to grasp. I remember saying to myself, "Of course I am what I think and feel. Like right now, I think this is not true, so I feel frustrated, and that's who I am!"

But as I gained more experience in meditation, I started to be able to observe my thoughts like a detached outsider might. So, if I'm the observer of my thoughts, feelings, and emotions, who is the person experiencing them? If I'm observing the observer, who is the observer? If I'm able to develop happy thoughts, and these thoughts create happy feelings and emotions, how can I be the *creator* of these feelings and emotions, and *be* those feelings and emotions at the same time? The answer is: I can't. In fact, I'm the only constant, as thoughts, feelings, and emotions change constantly.

Once I realized this, I started being able to analyze my thoughts exactly like I would someone else's. *This level of detachment is extremely liberating.* Because your thoughts, feelings, and emotions are not a part of who you are anymore.

They stop defining you. So not only can you drop them when needed, but there's no reason to be offended when they're challenged or criticized. You can be in charge of how you feel, which is your reality, and your own life. Is it easy? Not always, but like anything else, practice makes perfect. All of these useful understandings, I owe fully to the practice of meditation and to the teachings of Buddha.

In short, Buddha, like Socrates and the other sages, teaches us that happiness is an internal journey and can't be found through the altering of external factors. As we've discussed, no material possession can bring internal peace, rather, it's by way of a trained mind that this goal becomes attainable.

8.4 — Taoism

The philosophy, religion, or way of life, known as Taoism is believed to have been put forward by a mysterious man, Lao Tzu,[35] who lived in China in the 6th century BC.

The book he wrote, *Tao Te Ching*, which means the book of virtue, or the book of the Tao, the virtuous way, was intended for rulers of his time, as he was known to be a sage or a master. As I'm not an expert in this philosophy, I'll just share what I've found to be most useful, and will leave you to do more research if you so choose.

In some ways, Taoism is extra-difficult for Westerners to grasp, although its symbol (the Ying and the Yang), is familiar to almost everyone. Like a true Taoist dilemma, no one understands it, and yet everyone does. It's the most iconic of all Taoist symbols. It conveys that in all good, there is some bad, and in all bad, there is some good. Nothing is purely black and white. Everything carries some qualities of its opposite.

Taoism is a difficult philosophy to grasp for a mind that's been conditioned by Western thought, as its language and propositions tend to go against our common understanding of the way things are. It's full of seemingly-contradictory ideas, and that's precisely how it manages to offer a new window into the human condition. For example, take a look at these very first lines of the Tao Te Ching:

The Tao which can be expressed in words

is not the eternal Tao;

the name of the Tao which can be uttered,

is not its eternal name.

If this is true, that anything that could be said about it would not be accurate, why did Lao Tzu write quite a substantial book about it? I think the answer is two-fold:

1) As is also with the work of Rumi, Lao Tzu's writings are full of poetic inferences, which inspire imagination on the part of the reader. Without this structure, the Tao cannot be understood.

2) The truth is in actions, not in words. Maybe that's why, one of the meanings of Tao, is the way, of being or living. But the written word is of course, still necessary to point us toward the right action.

Taoism also goes against nearly all our Western beliefs. For instance, we're told that if we work hard and are honest, all our dreams will come true. Lao Tzu tells us the opposite, that we should not force anything, which is known as the Wu Wei, which means inaction or effortless action. He advises to work the right amount, at the right time, not to work hard.

His most notable teaching is that our actions should stand as the definition of who we are, not our words.

How does Taoism relate to financial stability? The three most important virtues of Taoism are *simplicity, patience, and compassion.* If you live simply and are patient with the way you handle your finances, you'll be more than half the way to a better-looking balance sheet.

Let's return to the idea of not forcing anything, Wu Wei. When you practice this, you eat only when you're hungry, and you stop when you're full. You work when you have the energy and motivation to do so, and rest when you're tired. You go to bed when you're sleepy, and get up when you're rested. You talk when you have something important to say, and listen when you don't. You spend quality time with the people important to you when that feeds your needs for social life, and seek alone-time when you need to reflect. When you follow this principle, all your actions become *fully engaged*, satisfactory, and meaningful, without excess. This practice can open the door to eating less, losing weight, becoming healthier, exercising just enough, being productive, improving everything you do, and developing more meaningful relationships. It can be a huge step toward embracing a simpler, happier, and more meaningful life.

As you read this, it's understandable that you might think that, since you work at a 9 to 5 job, or because your lunch hour is set, or because you have to get up at a certain time, that none of the above suggestions would be practical for you. That could of course be true. If so, consider the practice of these ideas when you can, and see if it makes a difference for you. For instance, while working on a project at home, and you have a chance to test it, try to not force anything, and see if you see a positive change in the outcome and your experience of the process. Our minds work better when relaxed, and forcing an issue can prove to be counterproductive.

With the idea of inaction, Wu Wei complements Buddhism, which suggests that one can not find happiness, but can allow it to be. In a non-dualistic fashion, being one with the source means we're naturally conditioned to be happy. It's our ego and the illusion of time, as Einstein put it, that is in the way. So the idea is that if we don't necessarily do anything, but if we let go of illusions, what will naturally emerge is happiness. You can't stop a wave in water, your very action to do so will create new waves. The only way to stop a wave is by inaction. That's Wu Wei, and that's also how you can find the self and happiness, according to Buddhism.

The second wisdom of Taoism that I find worth noting is the focus upon *actions and contemplation, as opposed to words*. It simply states that the right way, the Tao, can only be understood in action, and not by talking about it. Take small steps if need be. Start simplifying your kitchen and bedroom closet for instance, and you're on your way to the virtuous way. That's all you need, to be on the right path. When you're inclined to buy the next pair of shoes, take a look at what you already have, and if that's enough, don't force it, be patient. When you're tired, sit down and remain silent, close your eyes, and meditate on good thoughts. Your actions to simplify your life, to focus on being compassionate and patient, will put you on the right path, the virtuous path, the Tao.

Like Socrates, Lao Tzu teaches us to appreciate what we already have, and that being able to recognize what is enough, is true wealth. If we took the advice of these sages, we wouldn't have an overspending problem. We'd live simply, prudently, patiently, and with compassion. This is the Taoist way.

Conquering others takes force

Conquering yourself is true strength

Knowing what is enough is wealth

8.5 — Stoicism

"True happiness is to enjoy the present, without anxious dependence upon the future, not to amuse ourselves with either hopes or fears, but the rest, satisfied with what we have, which is sufficient, for he that is so wants nothing."

"We suffer more often in imagination than in reality. You want to live but do you know how to? You are scared of dying but tell me, is the kind of life you lead really any different than being dead?"

"Man is affected not by events, but by his views about them."

"If you really want to escape the things that harass you, what you're needing is not to be in a different place, but to be a different person."

"It is not the man who has too little, but the man who craves more, that is poor."

The quotes above are attributed to Seneca (4 BC-65 AD)

Seneca was born in Spain, which at that time was a part of the Roman Empire. As a sage, philosopher, teacher, and mentor, he rose to the top and became the tutor of Emperor Nero. Eventually it was Nero himself who ordered Seneca's death, forcing him to commit suicide, over what is considered to be a mistaken allegation that he'd had a role in a failed assassination plot against the Emperor. Seneca is probably the best-known philosopher of Stoicism, and he's the source of some of the most pertinent quotes and most actionable teachings for the purposes of this book.

Stoicism and Buddhism have a lot in common. According to Buddhism, a person who has reached Nirvana is one who can see things as they are. Similarly, Stoicism is a philosophy that teaches us that the path to happiness, is found in the acceptance of life as it is, and by not allowing ourselves to be influenced or controlled by our desires, emotions, and fears.

According to this school of philosophy, *virtue is all that's needed in order to be happy.* Virtue is, as mentioned earlier, the ability to find happiness in doing what's right.

Stoics believed that the answer to all sorts of questions could be found in nature, and if a person lived according to a natural order, that would be the good life. Stoics also put an emphasis upon the importance of *actions*, as the Taoists also do.

The name is derived from the Greek word "stoa," which means a porch, as the tradition's first teacher, Zeno, couldn't afford a building so he gave his teachings from a porch in Athens.

As an extremely progressive and revolutionary philosophy, the main ideas of stoicism were centered upon the ethics of living a life of self-control, guided by logic, and in accordance with natural laws. The stoics put forward the idea that all men were equal, including slaves, (which did lead to stoicism's popularity with enslaved peoples through the ages).

The three points stoics list for living a good life include:

1) Taking responsibility for your actions.

2) Living in the present, as opposed to in the past or in the future.

3) Focus upon what you can control.

This third point naturally supports the second, in that you can't change the past, and you have very little control over the future.

Again, similar to Buddhist teachings, stoics aimed to control the mind in order to find inner peace, regardless of external factors. They reasoned that if we allow what happens outside of us (those things which we can't control) to alter our emotions, then we become fully reactionary, and at the mercy of others' actions.

Unlike what many of us are conditioned to do, when an adverse event occurs, instead of saying "It will be OK," a stoic would say "It may not be OK, but that's not the end of the World, because I am bigger than my problems. I'll be a bigger and better person as a result of this experience."

Many of us are taught to think positively, which in and of itself may not be a bad thing, but this mental conditioning can cause problems if not balanced with some healthy skepticism and realism.

For instance, if you have to be at the airport at 2:00 pm, and it takes an hour to get there, a "think positive" person may not consider the myriad things that could go wrong, such as heavy traffic or long lines at the airport. Conversely, a stoic would contemplate all the things that could go wrong and would look for solutions to mitigate them. By doing so, if there happens to be an accident, they won't have to buy a new ticket, make changes to their travel plans, and thus end up having to spend more money than they anticipated. See how these ideas can all come together to help you stabilize your finances?

Perhaps, one of the best-known stoics in history is the Roman Emperor Marcus Aurelius (121-180 CE). In Ridley Scott's movie Gladiator (2000), there's a scene in which Marcus

Aurelius informs his son Commodus, that he wouldn't be taking his father's seat, to become the next Roman emperor. Understandably, Commodus gets very upset and tries to use reason in hopes of changing his father's mind. The answer he gives to his son, says it all about stoicism. He goes down on his knees and says apologetically: "Your faults as a son, are my failures as a father." He offered no sugarcoating of his conviction that his son lacked what it took to be the next emperor, and instead he takes full responsibility as the father. He doesn't say "You'll be OK." If he did, of course, he wouldn't be the Marcus Aurelius we're reading about today.

In short, stoics teach us to have the mental toughness to find happiness in living simply, with virtue, and to face our troubles head-on. This, along with taking responsibility for our actions will help us get the results we're after, especially when those results are put in the context of *spending rationally, saving carefully, and investing wisely*.

To get control of our lives and ultimately also our finances, — and to be happy and live a good life along the way — all we have to do is to look at the wisdom passed on to us from more than 2000 years ago and try to internalize these teachings.

Note that none of the sages advise us to chase luxury, live beyond our means, join the rat race, or work until we drop, but rather: live simply and consciously, be mindful in the present, and find happiness in virtue. Today, the strength of this advice is reinforced by what we know about the brain's biochemistry. It's up to us to execute based on what we've learned.

9
Conclusion

In my experience as a financial advisor, and based on plenty of research (which shows that saving and investing for future goals is an extremely difficult task in modern life), to simply tell people to start budgeting falls way short of solving the problem. Budgeting is only a small part of the solution. It's a complex issue, and the way out must cast a wider net to address all the related challenges. Some of the root causes are below.

- The biochemistry of our thoughts, feelings, and emotions, with their cause and effect feed-back loops have the potential to work against us. This vulnerability is masterfully exploited by sales techniques.

- Our cultural, economic, and political system is predominantly materialistic and capitalistic, and as a result, focuses more on corporate profits than an individual's wellbeing.

- It's difficult to navigate the complexity of our financial system, and as a result, those who can't afford to work with knowledgeable accountants, lawyers, and financial advisors risk falling behind. Because of the compounding effect, a period of 5 or 10 years of building debt with high interest rates can have devastating financial consequences.

   – Monotheistic religions lay the foundation for a
     judgmental belief system, and a dualistic way of
     looking at life, which can make it more difficult to find
     inner peace, a sense of connection to others, and
     natural happiness. This can open the door to
     either psychological maladies such as addictive
     personality disorder, or external self-medication
     strategies such as using drugs or overspending.

As a result, the solution will do best to be inspired by ideas from
these five areas: theology/philosophy, biology, psychology,
political science, and finance. It turns out that seemingly
unrelated pieces of information end up working well together
to answer our questions about overcoming the obstacles to
cultivating a healthy approach to personal finance, hence the
name of my theory on which this book is based: the Integrated
Wealth Theory.™

Admittedly, I've eclectically relayed information from topics
in which I have no professional expertise. This is true for all
the topics listed in the paragraph above, except of course for
the topic of personal finance. Experts in these fields may find
specific chapters of this book incomplete, and I respect that.
To those, I'd simply say: thanks for reading my book and let
me know how it can be improved. It's my hope to begin an
ongoing, open-minded conversation rather than offer a one-
time-only prognosis.

To those readers who found some parts of this book as
offensive, due to its political and/or religious content, I'd also
like to say that: if you're content with your beliefs, rather than
focusing upon that which you disagree, please focus on what
you do agree with, and use the information presented in these
parts of the book for your own benefit. There's nothing wrong
with integrating your beliefs into your finances, as long as
the result will carry you to success. I hope we can at least

agree upon the notion that a holistic, integrated approach is needed for success. I personally find no political or religious idea so holy that it's untouchable. There are useful and practical aspects to all ideologies and religions. I hope this collection of ideas will inspire you to organize your beliefs for the enhancement of your own wellbeing, while feeling free to pick and choose from them in any way that works for you.

In attempting to answer the question, "why is it that most people can't save money," we dissected the root causes of financially destructive behaviors. We found that since we are to some extent the products of our environment, then the political and socio-economic systems we live in deserve some criticism. But we can't stop there. On top of our action protocol hierarchy, for most people, consciously or unconsciously, their religious beliefs play an important role. It is one of the structural pillars of our culture. The axioms of religious/spiritual dogma need to be put on trial as well.

To add insult to injury, our brain structure reveals that biochemically and structurally, our thought processes and actions can simply be destructive. This is either due to the mechanics of our thoughts, or due to the chemicals released in the process of thinking them.

So, you have to be aware of all of the contributing factors, *and* learn at least the basics of personal finance, in order to find financial freedom. Stop Overspending and the proposed IWT, aims to be a road map for you on this journey. I hope you found the ideas presented here helpful.

I will finish my book with an ancient Egyptian myth, that teaches us one of Life's most valuable lessons: if you want a part of your life to improve, then pay attention to it. You will go where your attention is. You will in fact become what you pay attention to. So turn your attention to where you want to see some growth, expansion, or improvement. Since you picked up this book, I'm guessing you're seeking an improvement in your finances. Then pay attention to it. Pay attention to your income, and seek ways to increase it. Pay attention to your expenses, and seek ways to decrease them. Pay attention to your investments. Ultimately, pay attention to your life goals — if you do, you may find they have a way of coming to pass.

Leaving the best to last, here is our final story. I hope that you'll remember it for a long time: the Eye of Horus. This story explains the significance behind one practice of the Ancient Egyptians, their putting an eye on top of important buildings to signifying its societal stature. For them, the eye wasn't just an organ for seeing, but it also had powers of protection, healing and growth. Here we go...

Osiris was the God governing death and resurrection, and was the king of the Old Kingdom, which he built from the ground up. He was married to Isis, who was his sister and queen, and also a major goddess in Egyptian mythology, similar to the stature of Christianity's Virgin Mary. Isis had magic powers such as the ability to heal the sick, and to be a guide to the dead in the afterlife.

Osiris had a brother, Set. Set was jealous of his brother's position as King and had plans to overthrow him. As Osiris began to age and become blind, Set was conspiring to kill him and become the King.

One day Set, finding Osiris alone, slew him and chopped his body into pieces. He scattered his brother's body parts across the Kingdom, so Set would not feel threatened by him anymore. (As a god, if Osiris' pieces were found and put back together, he did have the power to resurrect. In fact, his spirit was not subject to death anyway).

Isis was devastated and wasn't going to stand by and watch her brother Set take Osiris' throne. She wandered Egypt looking for the pieces of Osiris, and being a goddess herself, she was able to find his reproductive organs. Together Osiris and Isis had a son named Horus.

Horus was half man half falcon. He was forced to grow up in exile, outside of the Old Kingdom that his father had built, where now his uncle, Set the Kingslayer sat on the throne. Horus looked forward to the day he could avenge his father.

Horus grew to be a strong young man. With a falcon's obvious visual powers, he could fly high in the sky and see with great acuity, the vast distances stretching ahead of him.

When he felt ready, Horus returned to the Old Kingdom and challenged his uncle Set. There was a hard-fought and bloody battle between the two. Horus was victorious but lost one of his eyes in the fight.

Horus could have become the New King, but instead, he traveled to the underworld, found his father, half dead and torn to pieces, and gifted him the eye he'd lost in battle with his uncle. Horus wanted Osiris to rule again, but this time, with focus. This turned out to be the most desirable combination: the vision of a young falcon and the wisdom of a wise king. Both qualities were needed for stability, because, just as Osiris couldn't be killed, so also was Set an immortal, capable of coming back to challenge the throne, and he would certainly bring chaos with him.

There's a lot of wisdom in this story, and its symbolism can be found in the myths of many cultures. The moral of the story is: you must pay attention to what you deem to be important, otherwise there will be chaos and a high price to pay. Having vision, keeping an eye on its progress, and protecting against threats to your progress, are your highest priorities for success.

Thanks for joining me on this journey, and best of luck to you!

Pay attention!

References

1   Bloomberg:

https://www.bloomberg.com/news/articles/2017-02-21/two-thirds-of-americans-aren-t-putting-money-in-their-401 k?sref=yR4bJbnb

2   GoBankingRates:
https://www.gobankingrates.com/retirement/planning/why-americans-will-retire-broke/

3   NY Federal Reserve Bank:
https://www.newyorkfed.org/microeconomics/hhdc.html

4   World Bank Data:
https://data.worldbank.org/indicator/NY.GDP.MKTP.CD?locations=US

5   Harry Stack Sullivan, Interpersonal Theory and Psychoterapy

6   Mind, Self and Society, Herbert Mead's students' notes

7   Jean-Jacques Rousseau, On the Social Contract, (Dover Thrift Editions)

8   Steven M Lukes, Individualism

9   Ludwig von Mises, Liberty and Property

9   Ludwig von Mises, The Ultimate Foundation of Economic Science, Theory and History, Human Action

10   Friedrich Nietzsche, Thus Spoke Zarathustra

11   Bernard Mandeville, The Fable of the Bees

12   Edward L. Bernays, Propaganda

13   Sigmund Freud, Civilization and Its Discontent

14   Joseph Campbell, The Hero's Journey

15   https://plato.stanford.edu/entries/epicurus/

16   https://www.pewresearch.org/fact-tank/2018/08/07/for-most-us-workers-real-wages-have-barely-budged-for-decades/

17   https://commons.wikimedia.org/wiki/File:Cumulative_percent_change_in_real_annual_earnings,_by_earnings_group,_1979-2017.png

18   Max Weber, The Protestant Ethic and the Spirit of Capitalism

19   Leonard Shlain, The Alphabet Versus the Goddess

20   Nouriel Harari, Home Sapiens

21   Nature Neuroscience, https://www.nature.com/articles/nn.4468

22   Marshmallow Test, Walter Mischel and Ebbe B. Ebbesen

23   Paul D. MacLean, The Triune Brain in Evolution

24,25   Robert Shiller, Behavioral Finance

24,25   Robert Shiler, Advances in Behavioral Finance

26,27   Dan Ariely, Predictably Irrational

28   Daniel Kahneman, Thinking Fast and Slow

29   Harvard Health Publishing, The Power of Placebo Effect, May 2017: https://www.health.harvard.edu/mental-health/the-power-of-the-placebo-effect

30   You Gov 2015 Survey: https://yougov.co.uk/topics/politics/articles-reports/2015/07/03/8-of-Britons-believe-horoscopes-predict-the-future

31   Britannica on Barnum Effect https://www.britannica.com/science/Barnum-Effect

32   Reminiscences of a Stock Operator, Edwin Lefèvre

33   Nicomachean Ethics, Aristotle

34   Socrates, https://www.pursuit-of-happiness.org/history-of-happiness/socrates/

35   Tao Te Ching, Lao Tzu

36   Erich Fromm, To Have or To Be

37   Francis Fukuyama, End of History and the Last Man